Modern Wiring Practice

Modern Wiring Practice

Design and Installation
Revised edition

W. E. Steward
and
T. A. Stubbs
With technical contributions by
Trevor Marks and Steve Clarke

ELSEVIER

Newnes

AMSTERDAM • BOSTON • HEIDELBERG • LONDON • NEW YORK • OXFORD
PARIS • SAN DIEGO • SAN FRANCISCO • SINGAPORE • SYDNEY • TOKYO
Newnes is an imprint of Elsevier

Newnes is an imprint of Elsevier
Linacre House, Jordan Hill, Oxford OX2 8DP, UK
30 Corporate Drive, Suite 400, Burlington, MA 01803, USA

First edition published 1952
Twelfth edition published 1995
Reprinted 1996, 1997, 1998, 2000, 2002 (twice), 2003
Revised and updated 2005
Reprinted 2006

British Library Cataloguing in Publication Data
Steward, W. E.
  Modern Wiring Practice – 12 Rev. ed.
  I. Title   II. Stubbs, T. A.
  621.31924

Library of Congress Cataloging-in-Publication Data
Steward, W. E. (William E.)
  Modern wiring practice: design and installation/W. E. Steward
  and T. A. Stubbs; with technical contributions by Trevor Marks
  and Steve Clarke – 13th ed.
  p. cm.
  includes index.
  ISBN 0 7506 6662 5 (pbk.)
  1. Electric wiring, Interior.   I. Stubbs, T. A.   II. Title.
  TK3271.S8 1994   94-24608
  621.319'24–dc20   CIP

ISBN–13: 978-0-7506-6662-6
ISBN–10: 0-7506-6662-5

For information on all Newnes publications
visit our website at www.newnespress.com

Printed and bound in *Great Britain*

06 07 08 09 10   10 9 8 7 6 5 4 3 2

Working together to grow
libraries in developing countries

www.elsevier.com | www.bookaid.org | www.sabre.org

ELSEVIER   BOOK AID International   Sabre Foundation

Contents

Northamptonshire
Libraries & Information
Services

Askews		

Preface

This book surveys the broad spectrum of electrical design and installation work, and this edition has been revised to incorporate the latest amendments to BS 7671 (The IEE Wiring Regulations) issued in 2004. The book is intended to supplement the various regulations and items of legislation. It is not a replacement for them.

The book is divided into two sections: (1) design of electrical installation systems and (2) practical work. The design section attempts to explain in simple terms the various regulations and requirements and goes on to deal with such matters as main switchgear, distribution, final circuits and special types of installations.

The practical sections, dealing with the most important wiring systems, is based on the author's experience, and includes many on-site illustrations and diagrams. The author hopes that readers will gain much useful information from the book. Any comments on the new edition will be most welcome.

T.A. Stubbs

Acknowledgements

I am grateful to many people for assistance with the preparation of this work: firstly, to the Institution of Electrical Engineers for much helpful advice, and for permission to publish extracts from the Wiring Regulations. The Regulations are published as a British Standard, BS 7671, and I am equally indebted to the BSI for their permission to publish extracts. This book is not a replacement for the IEE Regulations, and copies of these and the guidance notes which accompany them may be obtained from the Institution at PO Box 96, Stevenage, SGI 2SD.

I am also indebted to the British Standards Institution for authority to publish Figures 6.5 and 6.7. These are from BS 7375, and are reproduced with the permission of the BSI. Complete copies of the standard can be obtained by post from BSI Customer Services, 389 Chiswick High Road, London, W4 4AL.

Many individuals in the field of electrical design and installation work have been instrumental in giving advice which has helped me in the preparation of this edition. My numerous questions have been answered fully and courteously and this help has enabled me to present a practical and up-to-date volume. Many of the on-site photographs have been possible thanks to the agreement of individual electricians and designers, to whom I am most grateful.

My thanks also extend to the many electrical equipment suppliers who have provided illustrations: these are individually credited.

Finally I thank my wife for the many hours spent at the keyboard working on the text and, in the process, enhancing the quality of the English used in the book.

To one and all, I extend my appreciation and thanks.

Part 1
Design of Electrical
Installation Systems

1 Regulations governing electrical installations

Whatever type of electrical equipment is installed, it has to be connected by means of cables and other types of conductors, and controlled by suitable switchgear. This is the work which is undertaken by the installation engineer, and no equipment, however simple or elaborate, can be used with safety unless this installation work has been carried out correctly.

In the very early days of electricity there was no serious objection to running a pair of wires from a d.c. generator and connecting those wires by means of soldered T joints to branch circuits to which lamps and other current consuming appliances were connected. This load was continually added to until the generator produced more sparks than usual! If the cables got warm they were replaced by larger ones!

Planning of installation work

There was very little planning of wiring installations in those early days, but now, with supplies from the grid, very large sources of power are introduced into all premises which use electricity, and proper planning and design have become essential.

Like fire, electricity is a very good servant, but if not properly controlled and used it can prove to be a very dangerous master. The need for planned methods of wiring and installation work has long been recognised and all kinds of regulations, requirements, recommendations, codes of practice and so on have been issued. Some are mandatory and can be enforced by law, whilst others are merely recommendations.

As this book deals with the work of the installation engineer an attempt will be made to present, as clearly as possible, a general outline of the basis of good installation work, including planning and execution. References will be made to the various rules and regulations, and copies of these should be obtained and studied.

From what has already been said it should be clear to everyone who intends to undertake any electrical installation work that they must be conversant with all of the recognised standards and practices.

If an uninstructed amateur attempts to paint his house, at the very worst he can make an unsightly mess, but if he decides to install a few additional 'points' in his house, his workmanship might become a positive danger to himself and his family. Yet many people do undertake this work light heartedly and are sometimes encouraged to do so.

When planning an installation there are many things which must be taken into account: the correct sizes of cables, suitable switchgear, current rating of overcurrent devices, the number of outlets which may be connected to a circuit and so on. These and other matters are dealt with and explained in other chapters.

The regulations governing installation work can be divided into two categories: statutory regulations and non-statutory regulations.

Statutory regulations include:

Type of installation	Regulation	Administered by
Installations in general with certain exceptions	Electricity Safety, Quality and Continuity Regulations 2002	Secretary of State
All installations in the workplace including factories and offices	Electricity at Work Regulations 1989	Health and Safety Commission
Agriculture and horticultural installations	The Agricultural (Stationary Machinery) Regulations 1959	Health and Safety Commission
Electrical Equipment	The Low Voltage Equipment (Safety) Regulations 1989	Department of Trade and Industry
Building in Scotland with certain exceptions	Building Standards (Scotland) Regulations 1990	Secretary of State for Scotland
Health and safety at work	Health and Safety at Work Act 1974	Health and Safety Commission

Readers are advised to obtain from HMSO copies of the following publication: *Memorandum of Guidance on the Electricity at Work Regulations 1989*, which is published in three parts: *General; Mines* and *Quarries*.

Non-statutory regulations include:

Type of installation	Regulation	Published
Installations in general (with certain exceptions)	Requirements for Electrical Installations. IEE Wiring Regulations Sixteenth Edition BS 7671: 2001	British Standards Institution and the Institution of Electrical Engineers
Installations on construction sites	BS 7375: 1996 BS 4363: 1998	British Standards Institution
Installations in explosive atmospheres	BS EN 50014: 1998 BS EN 60079: 2003	British Standards Institution
Emergency lighting of premises (other than cinemas and similar premises)	BS 5266: 1999	British Standards Institution
Fire detection and alarm systems in buildings	BS 5839: 2002 BS EN 54	British Standards Institution
Protection of structures against lightning	BS 6651: 1999	British Standards Institution
Earthing	BS 7430: 1998	British Standards Institution

The Electricity Safety, Quality and Continuity Regulations 2002

The Electricity Safety, Quality and Continuity Regulations 2002 came into effect on 31 January 2003 and were drawn up with the object of securing a proper supply of electrical energy and the safety of the public. These new regulations replace The Electricity Supply Regulations 1988 and subsequent amendments up to and including those issued in 1998.

The Regulations apply to all 'duty holders' concerned with the supply and use of electrical energy and these include generators, distributors, transmitters, meter operators and others supplying electricity to consumers. They also apply to the agents, contractors and subcontractors of any duty holders.

As with the earlier regulations, parts of the 2002 regulations apply to the supply of electricity to consumer's installations (Regulations 23 to 29 inclusive) and give the electricity distributor powers to require certain standards of installation before giving or maintaining a supply to the consumer. Regulation 25(2) states that 'A distributor shall not give his consent to the making or altering of the connection where he has reasonable grounds for believing that the consumer's installation fails to comply with British Standard Requirements.'

If any installation is not up to standard, the distributor may issue a notice in writing to the consumer requiring remedial works to be carried out within a reasonable period. The period required must be stated in the notice. If remedial works are not carried out by the end of the period specified, the distributor may disconnect (or refuse to connect) the supply and, in the event of such disconnection must set out the reasons in a further written notice.

A distributor may also disconnect a supply without giving notice, if such disconnection can be justified on the grounds of safety. In this event the distributor must give notice in writing as soon as reasonably practicable, giving reasons and details of remedial measures required. The distributor shall restore the supply when the stipulated remedial measures have been taken.

If there is a dispute between the distributor and consumer over the disconnection or refusal to connect, which cannot be resolved between them, the matter may be referred to the Secretary of State who shall appoint a suitably qualified person to determine the dispute. Following the determination, the distributor shall maintain, connect, restore or may disconnect the supply as appropriate, subject to any conditions specified in the determination.

IEE Wiring Regulations – BS 7671

The full title is 'Requirements for electrical installations – The IEE Wiring Regulations – Sixteenth Edition. BS 7671: 2001, and is based upon CENELEC (The European Committee for Electrotechnical Standardisation) Harmonisation Documents formed from IEC (International Electrotechnical Commission) standards. The requirements and some of the actual wording are therefore similar to IEC standards.

The IEE Regulations are divided into the following parts:

Part 1   Scope, object and fundamental principles
Part 2   Definitions
Part 3   Assessment of general characteristics
Part 4   Protection for safety
Part 5   Selection and erection of equipment

Part 6   Special installations or locations
Part 7   Inspection and testing.

There are also 6 appendices, and these are:

Appendix 1   *British standards to which reference is made in the Regulations*
Appendix 2   *Statutory regulations and associated memoranda*
Appendix 3   *Time/current characteristics of overcurrent protective devices*
Tabular and graphical data is included for both fuses and miniature circuit breakers. Fuses to BS 88, BS 1361 and BS 3036 are included as well as mcb types B, C and D (BS EN 60898 and BS EN 61009).
Appendix 4   *Current-carrying capacity and voltage drop for cables and flexible cords*
Tables are included for cables with copper or aluminium conductors.
Appendix 5   *Classification of external influences*
This gives a list of external influences which must be taken into account when designing the installation. The coding system is that used in International Electrotechnical Commission (IEC) publication 364.
Appendix 6   *Model forms for certification and reporting*
Appendix 7   *Harmonized cable core colours.*

In addition to the Regulations themselves, the IEE also publish books of Guidance Notes and an *On-site guide.* The subjects covered include:

Protection against fire
Protection against electric shock
Protection against overcurrents
Isolation and switching
Selection and erection of equipment
Testing and inspection
Special installations and locations
Design procedure with design data.

The books provide much additional useful information over and above that contained in the 16th edition of the Wiring Regulations themselves.
The present book is based upon the requirements of the 16th edition of the IEE Regulations, and the following comments on each

part are offered for the benefit of readers who are not familiar with
the layout and presentation.

Part 1 Scope

The scope of the Regulations relates to the design, selection and erection of electrical installations in and about buildings. The Regulations cover the voltage up to and including 1000 V a.c. or 1500 V d.c.

They also cover certain installations exceeding this voltage, for
example, discharge lighting and electrode boilers.

The Regulations do not apply to electrical equipment on ships, off-
shore installations, aircraft, railway traction equipment, motor
vehicles (except caravans) or to the aspects of mines and quarries
which are specifically covered by Statutory Regulations.

Object   The Regulations are intended to provide for the safety of
persons, property and livestock, against dangers and damage which
may arise during reasonable use of the installation.

The fundamental principles of the Statutory Regulations are satisfied
if the installation complies with Chapter 13 of the IEE Regulations.

Fundamental requirements for safety   The fundamental
requirements enumerated in Chapter 13 of the IEE Regulations form
the basis on which the remainder of the Regulations are built. This
fundamental requirement is also used in the Electricity Safety Regulations and the Electricity Regulations of the Factories Act, but in
slightly different words.

Two aspects which are included in the fundamental requirements
are worthy of emphasis. Safety does depend upon the provision of a
sound, well thought out, electrical design, and also the expertise of
good electricians doing a good, sound job. This latter requirement is
expressed in IEE Regulation 133-01-01 which states: 'Good workmanship and proper materials shall be used'. Another item worthy
of note states that equipment shall be arranged so as to afford sufficient space for installation and accessibility for operation, inspection,
testing, maintenance and repair.

Alterations to installations   This aspect is worthy of special com-
ment as there are significant implications in the requirements. The
subject is covered in IEE Regulations 130-07 and in Section 743. Any
alterations to an existing installation must, of course, comply with the
IEE Wiring Regulations, and this includes any part of the existing
work which becomes part of the alteration. In addition the person

making the alteration must ensure that the existing arrangements are capable of feeding the new part safely. This in practice means that the existing installation must be subject to tests to ascertain its condition. It is not the duty of the installer to correct defects in another part of the system, but it is his duty to advise the person ordering the work. This advice should be in writing. In practice it may be preferable to start the altered wiring from a new distribution board.

Part 2 Definitions

A comprehensive list of definitions used in the IEE Regulations is contained in Part 2 of the Regs. These definitions will occur constantly and a clear understanding is necessary in order to plan and execute installations. Some of the terms are given below.

Protective conductor A conductor used for some measures of protection against electric shock and intended for connecting together any of the following parts: exposed conductive parts, extraneous conductive parts, the main earthing terminal, earth electrode(s), the earthed point of the source, or an artificial neutral.

Earthing conductor A protective conductor connecting the main earthing terminal of an installation to an earth electrode or to other means of earthing.

Equipotential bonding Electrical connection maintaining various exposed conductive parts and extraneous conductive parts at substantially the same potential.

Functional earthing Connection to earth necessary for the proper functioning of electrical equipment.

PEN conductor A conductor combining the functions of both protective conductor and neutral conductor.

Circuit protective conductor (cpc) A protective conductor connecting exposed conductive parts of equipment to the main earth terminal.

Live part A conductor or conductive part intended to be energised in normal use, including a neutral conductor but, by convention, not a PEN conductor.

Barrier A part providing a defined degree of protection against contact with live parts, from any usual direction of access.

Bunched Two or more cables contained in a single conduit, duct, ducting, or trunking or, if not enclosed, are not separated from each other by a specified distance.

Overcurrent   A current exceeding the rated value. For conductors the rated value is the current-carrying capacity.

Circuit breaker   A device capable of making, carrying and breaking normal load currents and also making and automatically breaking, under pre-determined conditions, abnormal currents such as short circuit currents. It is usually required to operate infrequently although some types are suitable for frequent operation.

Residual current device   A mechanical switching device or association of devices intended to cause the opening of the contacts when the residual current attains a given value under specified conditions.

Exposed conductive part   A conductive part of equipment which can be touched and which is not a live part but which may become live under fault conditions (e.g. conduit, trunking, metal enclosures, etc.).

Extraneous conductive part   A conductive part liable to introduce a potential, generally earth potential, and not forming part of the electrical installation.

SELV (separated extra-low voltage)   An extra-low voltage system which is electrically separated from earth and from other systems in such a way that a single fault cannot give rise to the risk of electric shock.

PELV (protective extra-low voltage)   An extra-low voltage system which is not separated from earth, but which otherwise satisfies all the requirements for SELV.

Direct contact   Contact of persons or livestock with live parts.

Indirect contact   Contact of persons or livestock with exposed conductive parts which have become live under fault conditions.

Part 3 Assessment of general characteristics
Chapters 31, 32, 33 and 34 of the Regulations firmly place responsibility upon the designer of the installation to ensure that all relevant circumstances are taken into account at the design stage. These considerations include the following characteristics:

(1) Maximum demand
(2) Arrangements of live conductors and type of earthing
(3) Nature of supply
(4) Installation circuit arrangements.

Part 4 Protection for safety

This section covers:

Protection against electric shock
Protection against thermal effects, e.g. fire and burns and overheating
Protection against overcurrent
Protection against overvoltage
Protection against undervoltage
Isolation and switching
Application of protective measures for safety
Choice of protective measures as a function of external influences.

These matters are dealt with in detail in Part 4 of the IEE Regulations, in Chapters 41, 42, 43, 44, 45, 46, 47 and 48 respectively.

Part 5 Selection and erection of equipment

This section covers:

Common rules, such as compliance with standards
Selection and erection of wiring systems
Switchgear
Earthing and protective conductors
Other equipment such as transformers, rotating machines, etc.
Supplies for safety services.

These matters are dealt with in detail in Chapters 51–56 of the IEE Regulations.

Part 6 Special installations or locations

The 16th edition of the IEE Wiring Regulations introduced a new section for special types of installation. The Regulations give particular requirements for the installations and locations referred to, and these supplement or modify the requirements contained in other parts of the Regulations.

Installations and locations covered include bath/shower rooms, swimming pools, saunas, construction sites, agricultural and horticultural premises, caravans and motor caravans, caravan parks and highway equipment (street lighting etc.). There are also regulations on conductive locations, and earthing requirements for equipment (such as data processing equipment) which has a protective conductor current exceeding 3.5 mA.

Part 7 Inspection and testing

The responsibility for inspection and testing demands a range of techniques and types of instrument. Full details of necessary tests are given in Chapters 71, 72, 73 and 74 of the IEE Regulations.

The Electricity at Work Regulations 1989

These Regulations came into force on 1st April 1990 and apply to all electrical systems installed in places of work. They are more wide ranging than the regulations they replace, as they now apply to all places of work, including shops, offices, etc., as well as factories, workshops, quarries and mines which were covered by previous legislation. They also relate to safety arising from any work activity – not just electrical work – being carried out either directly or indirectly on an electrical system, or near an electrical system.

The Regulations place duties upon all employers, self-employed persons, managers of mines and quarries and upon employees, and cover the construction, maintenance and work activities associated with electricity and electrical equipment. The Regulations come under the jurisdiction of the Health and Safety Commission.

A number of regulations have been revoked or modified as a result of the new legislation and these are listed in full in Schedule 2 of the Electricity at Work Regulations 1989. Some of the main ones are:

The Electricity Regulations 1908
The Electricity (Factories Act) Special Regulations 1944
The Coal and Other Mines (Electricity) Order 1956
The Miscellaneous Mines (Electricity) Order 1956
The Quarries (Electricity) Order 1956.

There are 33 regulations in the 1989 edition, and Regulations 4 to 16 apply to all installations and are general in nature. Regulations 17 to 28 apply to mines and quarries. Regulations 29 to 33 cover miscellaneous points. Three books are available from the HMSO which give additional information and guidance and it is recommended they be obtained and studied. Book 1 covers the Regulations in general, and the other two relate to mines and quarries respectively.

The Electricity at Work Regulations 1989 impose a number of new items and there is a change in emphasis in some regulations which significantly alter their application when compared with the regulations they replace. The paragraphs which follow give a brief description of some of the main features.

General No voltage limitations are specified, and the Regulations apply to all systems. Two levels of duty are imposed and these are (1) absolute and (2) as far as is reasonably practicable. The Regulations themselves indicate which level of duty applies to a particular regulation, and further help is given in the *Memorandum of Guidance*.

Regulations 1 to 3 Introduction These form the introduction, give definitions and state to whom the Regulations apply.

Regulation 4 General This is divided into four parts which cover (1) system design and construction, (2) system maintenance to ensure safety, (3) all work activities on or near the system and (4) provision of protective equipment for persons. All work activities are covered (not just electrical work) and this is sometimes referred to as the 'catch all' regulation. Three of the parts are to be implemented 'as far as is reasonably practicable', but the fourth, on the provision of protective equipment, is absolute. Note that in the definitions a system covers equipment which 'is, or may be' connected to an electrical supply.

Regulation 4(2) refers to system maintenance and it is intended that planned preventative maintenance is used and that the system design is such that this can take place. In this connection it should be noted that adequate working space must be provided. Further details are given under Regulation 15 below.

Regulation 5 Strength and capability Both thermal and mechanical provision are to be considered, and the arrangement must not give rise to danger *even* under overload conditions. Insulation, for example, must be able to withstand the applied voltage, and also any transient overvoltage which may occur.

Regulation 6 Environments This regulation relates to equipment exposed to hazardous environments, which can be mechanical damage, weather conditions, wet or corrosive atmospheres or from flammable or explosive dusts or gases. There is an important change when compared to the earlier regulations in that the exposure needs to be foreseen, knowing the nature of the activities undertaken at the site, and the environment concerned. This requires a degree of understanding between the designer and the user of the equipment.

Regulation 7 Insulation etc. Requires that conductors be suitably insulated and protected or have other precautions taken to prevent danger. A number of industrial applications will require precautions to be taken to suit the need, where provision of insulation is impractical.

For example, with conductor rails of an electrified railway, precautions may include warning notices, barriers or special training for the railway staff. As another example, the use of protective clothing is a requirement of use of electric welding equipment.

Regulation 8 Earthing   Requires earthing or other precautions to prevent danger from conductive parts (other than conductors) becoming charged. Metallic casings which could become live under a fault condition are included, and also non-metallic conductors such as electrolyte. Earthing and double insulation are the two most common methods of achieving the requirements, but six others are listed in the *Memorandum of Guidance*.

Regulation 9 Integrity   Intended to ensure that a circuit conductor connected to earth or other referenced conductors do not become open circuit or high impedance which could give rise to danger. Reference is made in the guidance notes both to combined and to separate neutral and protective conductive conductors.

Regulation 10 Connections   Must be sound, and suitable for purpose, whether in permanent or temporary installations. In particular, connections such as plugs and sockets to portable equipment need to be constructed to the appropriate standards. Also, where any equipment has been disconnected (e.g. for maintenance purposes) a check should be made as to the integrity of the connections before restoring the current, as loose connections may give rise to danger from heating or arcing.

Regulation 11 Excess current protection   It is recognised that faults may occur, and protection is needed usually in the form of fuses or circuit breakers to ensure that danger does not arise as a result of the fault. Every part of the system must be protected, but difficulties can arise since in fault conditions, when excess current occurs, it takes a finite time for the protective fuse or circuit breaker to operate. The 'Defence' Regulation 29 applies, and good design, commissioning and maintenance records are essential. The IEE Regulations give further guidance on this subject.

Regulation 12 Isolation   Requires provision of suitable means whereby the current can be switched off, and where appropriate, isolated. Isolation is designed to prevent inadvertent reconnection of equipment and a positive air gap is required. Proper labelling of switches is also needed. IEE Regulations 130-06 and 461 are relevant and are described on page 38 of this book.

Fig. 1.1 *To comply with the Electricity at Work and IEE Regulations, it is necessary, in appropriate circumstances, to provide means to 'prevent any equipment from being inadvertently or unintentionally energised'. Isolators with provision for padlocking in the isolated position are available to meet this requirement (Legrand Electric Ltd)*

Regulation 13 Working dead Precautions to prevent dead equipment from becoming live whilst it is being worked on are required, and can include the locking of isolators, removal of links, etc. Isolation, must obviously be from *all* points of supply, so it is a necessity for the operator to be familiar with the system concerned.

Regulation 14 Working live The intention is that no work on live conductors should be undertaken. However, it is recognised that in certain circumstances live working may be required, and the regulation specifies three conditions which must *all* be met before live working is to be considered. Care must be given to planning such an operation, and if live working is unavoidable, precautions must be taken which will prevent injury. It should be noted that the provision of an accompanying person is not insisted upon, and it is for consideration by those involved whether such provision would assist in preventing injury. If accompaniment is provided, the person concerned clearly needs to be competent. In cases where two equal grade persons work together, one of them should be defined as party leader.

Regulation 15 Access Requires that proper access, working space and lighting must be provided. In this connection the contents of Appendix 3 of the *Memorandum of Guidance* should be noted. This refers to legislation on working space and access, and quotes Regulation 17 (of the 1908 Regulations) which should be given proper consideration.

In this minimum heights and widths of passageways are specified to ensure safe access can be obtained to switchboards.

Regulation 16 Competence   The object of this regulation is to ensure that persons are not placed at risk due to lack of knowledge or experience by themselves or others. Staff newly appointed may have worked in quite different circumstances, and there is a duty to assess and record the knowledge and experience of individuals.

Regulations 17 to 28 Mines and quarries   These regulations apply to mines or quarries, and separate books of guidance are available from HMSO.

Regulation 29 Defence   Applies to specific regulations (which are listed in the Regulations) and provides that it shall be a defence (in criminal proceedings) to prove that all reasonable steps were taken in avoiding the commission of an offence. In applying this regulation it would be essential to maintain proper records and this is relevant for design, commissioning and maintenance matters. Also proper recording of design parameters and assumptions is necessary.

Regulation 30 Exemptions   No exemptions have been issued at the time of writing.

Regulations 31 to 33 General   These refer to application outside Great Britain, and to application to ships, hovercraft, aircraft and vehicles. Regulations revoked or modified are also listed.

British Standards

Since 1992 the IEE regulations themselves have been issued as a British Standard, BS 7671. In addition, there are many other British Standards which affect electrical installations, and these are designed to encourage good practice. These Standards go into more detail than the other regulations mentioned and IEE Appendix I lists those to which reference is made.

The Low Voltage Electrical Equipment (Safety) Regulations 1989

These regulations impose requirements relating to the safety of electrical equipment. They apply to equipment designed for use at a voltage not less than 50 V a.c. and not more than 1000 V a.c. (75–1500 V d.c.).

The Regulations are statutory and are enforceable by law. They are intended to provide additional safeguards to the consumer against accident and shock when handling electrical appliances. The main requirements are that equipment must be constructed in accordance with good engineering practice, as recognised by member states of the EEC. If no relevant harmonised standard exists, the Regulations state which alternative safety provisions apply.

The requirements state that equipment is to be designed and constructed so as to be safe when connected to an electricity supply and mechanical as well as electrical requirements are specified. If the user needs to be aware of characteristics relevant to the safe use of the equipment, the necessary information should if practicable be given in markings on the equipment, or in a notice accompanying the equipment. Other detailed information is given in the Regulations and in the explanatory notes.

Health and Safety at Work Act 1974 (reprinted 1977)

The three stages of this Act came into force in April 1975. It partially replaces and supplements the Factories Act, and the Offices, Shops and Railway Premises Act. It applies to all persons at work, whether employers, employees and self-employed, but excludes domestic servants in private households.

The Act covers a wide range of subjects, but as far as electrical installations are concerned its requirements are mainly covered by those of the Regulations for Electrical Installations, issued by The Institution of Electrical Engineers, and The Electricity at Work Regulations.

The Health and Safety Executive has issued booklets which give detailed suggestions on various aspects as to how to comply with these requirements. The booklets which mainly affect electrical installations are:

HS (G) 13 Electrical testing
HS (G) 38 Lighting at work
HS (G) 41 Petrol filling stations: construction and operation.

The main object of the Act is to create high standards of health and safety, and the responsibility lies both with employers and employees. Those responsible for the design of electrical installations should study the requirements of the Act to ensure that the installation complies with these.

2 *Designing an electrical installation*

Those responsible for the design of electrical installations, of whatever size, must obtain and study very carefully the requirements of the IEE Regulations for Electrical Installations, and also statutory regulations, details of which are given in Chapter 1.

The 16th edition of the IEE Regulations deals with the fundamental principles and gives the electrical designer a degree of freedom in the practical detailed arrangements to be adopted in any particular installation. It is necessary to be sure that the detailed design does in fact comply with the requirements laid down, and as a result a high level of responsibility has to be carried by those concerned with installation planning and design. In many cases, the experience and knowledge of the designer will be called into play to arrive at the best or most economical arrangement and this will encompass the practical application of installation techniques, as well as the ability to apply the theoretical aspects of the work. It will generally be necessary to demonstrate compliance with the Regulations and, in view of this, records need to be kept indicating the characteristics of the installation, the main design calculations and the assumptions made in finalising the design.

This chapter attempts to describe some of the points a designer will need to consider when planning an electrical installation. To enable the reader to refer to the relevant parts of the IEE Wiring Regulations more easily, each section of this chapter is headed by a guide to the relevant parts of the Regulations thus:

Part   Chapter   Section   Regulation   Appendix

Assessment of general characteristics

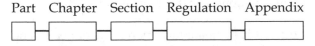

Part	Chapters	Section	Appendix
3	31 to 35	546	5

Before any detailed planning can be carried out, it is necessary to assess the characteristics of the proposed scheme. This applies whether the installation is a new one, an extension to an existing system or an electrical rewire in an existing building. The assessment required is a broad one and some of the aspects to be considered are described below.

The purpose and intended use of the building and the type of construction

The construction and use of the building will inevitably dictate the type of equipment to be installed. Environmental conditions, utilisation and construction of buildings are dealt with in IEE Appendix 5.

Environmental conditions include ambient temperature, altitude, presence of water, dust, corrosion, flora, fauna, electromagnetic or ionising influences, impact, vibration, solar radiation, lightning and wind hazards. Features in the design will depend upon whether the building is occupied by skilled technicians, children, infirm persons and so on, also whether they are likely to have frequent contact with conductive parts such as earthed metal, metal pipes, enclosures or conductive floors.

Fire risks in the construction of buildings are also covered in IEE Appendix 5. With large commercial premises, these matters may be dealt with by the fire officer, but the electrical designer needs to be aware of such factors as combustible or even explosive contents and also the means of exit. Supplies for safety services are covered in IEE Regulation 313-02.

The maximum current demand

It is necessary to estimate the maximum current demand. IEE Regulation 311-01-01 states that diversity may be taken into account. Diversity is not easy to define, but can be described as the likely current demand on a circuit, taking into account the fact that, in the worst possible case, less than the total load on that circuit will be applied at any one time. Additional information is given in the IEE *On-site guide*, Appendix 1.

An example may help to make this clear. With cooking appliances, not all the connected load is likely to be used at the same time. It may well be appropriate to apply diversity in a canteen or hotel situation, but if the cooking appliances are installed in a school domestic science room, it would be unwise to apply diversity as in practice there will be a tendency for all the appliances to be used simultaneously. Whilst the application of diversity is often an advantage in reducing the size of conductors and the associated protective devices, the application must be made with care. The designer is responsible for ensuring the conditions likely to prevail are appropriate ones for its use.

It would also be normal practice when at the planning stage, to make an allowance for future anticipated growth.

The characteristics of the supply including the type of earthing arrangements

With regard to the arrangement of live conductors, and the type of earthing arrangement, the electricity supplier should be consulted. It will be necessary to establish:

The nominal voltage
Nature of current and frequency
Prospective short circuit current at the origin of the installation
Type and rating of the supplier's overcurrent protective devices
Suitability of the supply
Earth loop impedance external to the installation.

If it has been decided to provide a separate feeder or system for safety or standby purposes, the electricity supplier shall be consulted as to the necessary changeover switching arrangements.

Installation circuit arrangements cover the need to divide circuits as necessary to avoid danger, minimise inconvenience and facilitate safe operation, inspection, testing and maintenance (IEE Regulation 314-01-01). Thus, for example, it is good practice to spread lighting between different circuits so as to prevent a circuit failure causing a complete blackout.

Compatibility of the equipment

The section under the heading of 'Compatibility' deals with characteristics which are likely to impair, or have harmful effects upon other electrical or electronic equipment or upon the supply. These include:

Possible transient overvoltages
Rapidly fluctuating loads
Starting currents
Harmonic currents (e.g. from fluorescent lighting)
Mutual inductance
d.c. feedback
High frequency oscillations
Earth leakage currents
Need for additional earth connections.

Fig. 2.1   *IEE Regulation 513 requires that every item of equipment be arranged so as to facilitate its operation, inspection and maintenance. This switchboard is installed in a room provided for the purpose, with access space both in front and behind the equipment. This situation will allow access for proper maintenance (W.T. Parker Ltd)*

Suitable isolating arrangements, separation of circuits or other installation features may need to be provided to enable compatibility to be achieved. Rapidly fluctuating loads, or heavy starting currents may arise in, for example, the case of lift motors or large refrigeration compressors which start up automatically at frequent intervals, causing momentary voltage drop. It is advisable for these loads to be supplied by separate submain cables from the main switchboard, and sometimes from a separate transformer.

Electromagnetic compatibility

Regulations regarding Electromagnetic Compatibility (EMC) came into effect in January 1996. These regulations (The Electromagnetic Compatibility Regulations: 1992) are derived from European Community directives and require that all electrical equipment marketed in the community meets certain standards. The aim of the regulations is to reduce the electromagnetic interference from equipment such that no harmful effects occur; also the equipment itself must be designed to have inherent immunity to any radiation it receives.

Provided equipment satisfies the requirements, the CE mark may be affixed by the manufacturer. Two methods are available to demonstrate compliance. These are the standards-based route, requiring tests by an independent body, or the construction-file route where the manufacturer records the design and the steps taken to achieve compliance. In practical terms the aspects which need to be taken into account include selection of components, earthing arrangements, equipment layout and design of circuits, filtering and shielding.

Installation engineers will be seeking to use equipment which carries the CE mark but it is important to bear in mind that installation practice itself has a bearing on the compatibility of equipment. It is possible that the installation arrangements may fail to preserve the inherent EMC characteristics of the individual items of equipment. With large installations or on complex sites, a systematic approach to EMC will be needed. Possible remedies could include the installation of additional screening or filtering, or require cables to be rerouted to address the situation.

A book such as this cannot address this complex subject in detail but specialist advice can be obtained from consultants in this field of work.

Maintainability

Consideration must be given to the frequency and to the quality of maintenance that the installation can reasonably be expected to receive.

Under the Electricity at Work Regulations 1989, proper maintenance provision must be made including adequate working space, access and lighting. The designer must ensure that periodical inspection, testing, and maintenance can be readily carried out and that the effectiveness of the protective measures and the reliability of the equipment are appropriate to the intended life of the installation. It is necessary to ensure as far as possible that all parts of the installation which may require maintenance remain accessible. Architects need to be aware of these requirements and in modern commercial buildings it is often the practice to provide special rooms for electrical apparatus.

In carrying out the design process, a number of decisions will be needed at the 'assessment of general characteristics' stage. It is important to record this data so that when required either during the design stage or afterwards, reference can be made to the original assessment process. The person carrying out final testing will require information as to these design decisions so as to assess the design concept employed in the installation.

Systems of supply

Chapters	Sections	Regulation
41 & 47	312, 412 & 413	312-03

A number of aspects of design will depend upon the system of supply in use at the location concerned. Five system types are detailed in the IEE Regulations. The initials used indicate: the earthing arrangement of the supply (first letter), the earthing arrangement of the installation (second letter), and the arrangement of neutral and protective conductors (third and fourth letters).

TN-S system A system (Fig. 2.2) having the neutral point of the source of energy directly earthed, the exposed conductive parts being connected to that point by protective conductors, there being separate neutral and protective conductors throughout the system. (This is the old system in Great Britain, which is gradually being changed over to a TN-C-S system.)

TN-C-S system As above but the neutral and protective conductors are combined in part of the system, usually the supply (PME supply) and then separate in the rest of the system (usually the installation).

TN-C system (Fig. 2.3) As above but the neutral and protective conductors are combined throughout the system.

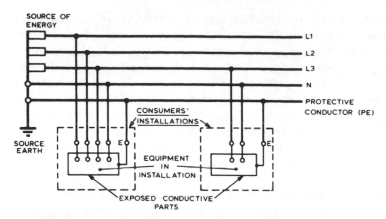

Fig. 2.2   *TN-S system. Separate neutral and protective conductors throughout system. The protective conductor (PE) is the metallic covering of the cable supplying the installations or a separate conductor. All exposed conductive parts of an installation are connected to this protective conductor via the main earthing terminal of the installation (E)*

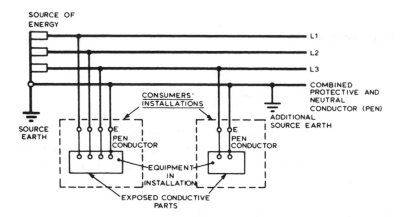

Fig. 2.3   *TN-C system. Neutral and protective functions combined in a single conductor throughout the system. All exposed conductive parts of an installation are connected to the PEN conductor. This system is only used in distribution networks as The Electricity Safety, Quality and Continuity Regulations prohibit combination of neutral and protective conductors in a consumer's installation*

TT system   (Fig. 2.4) One point of the source of energy directly earthed, but the exposed conductive parts of the installation being connected to earth electrodes independent of the earth electrodes of the power system.

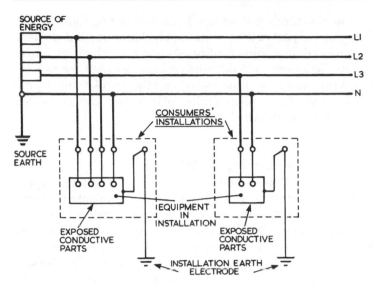

Fig. 2.4 *TT system. All exposed conductive parts of an installation are connected to an earth electrode which is electrically independent of the source earth*

IT system A system where the neutral point of the source of energy is either isolated from earth or connected to earth through a high impedance. The exposed conductive parts of the installation are earthed via an earth rod. The electricity companies are not allowed to use this system on the low voltage distribution network to the public.

Three systems are in general use in the United Kingdom at the present time. These are the TN-S, TN-C-S and TT types. The supply undertaking may well provide an earthing terminal at the consumer's installation, and this constitutes part of a TN system. In some cases the protective conductor is combined with the neutral conductor, and in this arrangement the system is the TN-C-S type, the supply being known as Protective Multiple Earthing (PME).

New systems coming into use nowadays are increasingly of the TN-C-S type, but before use can be made of the PME type of supply, stringent conditions must be met. Special requirements may apply and the supply undertaking must be consulted. If no earth terminal can be provided by the supply undertaking or the supply does not comply with the conditions for PME, then the TT system must be used.

Each of the systems described demands different design characteristics and the designer must take such factors into consideration when planning the installation.

The IT system is rarely encountered and will not be found in the United Kingdom as part of the public supply. However, it does have particular application in some continuous process industries, where an involuntary shutdown would cause difficulties with the process concerned. A private supply is necessary and the Regulations require a means of monitoring system faults so that they cannot be left undetected (IEE Regulation 413-02-24).

Protective multiple earthing (PME)

Although some installations today are to the TT system, supply undertakings are increasingly providing an earth terminal to the consumer, and as described above this may, on rare occasions, be to the TN-S system. However, PME supplies are now being applied more generally, and the tendency will be for new systems to be TN-C-S.

PME is defined as an earthing arrangement, found in TN-C-S systems, in which the supply neutral conductor is used to connect the earthing conductor of an installation with earth. The arrangement has some dangers but the main advantage of the system is that any earth fault which occurs, automatically becomes a phase to neutral fault, and the consequent low impedance will result in the fast operation of the protective devices.

Multiple earthing of the neutral is a feature of a PME supply and this is employed to ensure that in the event of a broken neutral, dangerous voltages do not occur. High standards of installation are applied by the supply undertaking to reduce the likelihood of an open circuit neutral conductor, and any installation connected to a PME supply must be to the same high standard. If the distributor offers a PME earthing terminal, it may only be used by the consumer if the installation complies with the requirements of the Electricity Safety, Quality and Continuity Regulations 2002. IEE Regulation 413-02-02 refers to bonding, and minimum sizes of supplementary bonding conductors are given in IEE Regulation 547-02. It is also necessary to look at Regulation 413-02-28 for sizing the supplementary bonds.

It should be noted that installation of PME is specifically prohibited in petrol filling stations under Health and Safety booklet HS (G) 41 'Petrol Filling Stations: Construction and Operation'. The reason being that the exposed conductive parts would have a voltage on them with respect to true earth and parallel earth paths causing high return currents to flow to earth through the petrol station equipment such as dispenser fuel pipes and underground storage tanks must be avoided.

Protection for safety

This is based upon Part 4 of the IEE Regulations. Chapters 41, 42, 43, 45 and 46 refer to different aspects of the topic and Chapter 47 covers the application of the measures listed in the Regulations. For ease of reference there is a relationship between the numbers, in that the application of Chapter 41 is covered in Section 471, Chapter 42 in Section 472, and so on. The areas covered are:

Protection against electric shock
Protection against thermal effects
Protection against overcurrent, both overload and short circuit
Protection against undervoltage
Isolation and switching.

Protection against electric shock

Chapters   Section

| 41, 53 & 54 |—| 471 |

The IEE definition of electric shock is 'A dangerous physiological effect resulting from the passage of an electric current through a human body or livestock.' The value of the shock current liable to cause injury depends on the circumstances and individuals concerned. Protection must be afforded in normal service and in the case of a fault. These are referred to in the IEE Regulations as protection against 'direct

Fig. 2.5 *An isolating transformer supplying extra low-voltage socket outlets for use of portable tools in an industrial installation (British Telecom)*

contact' and protection against 'indirect contact' respectively. A number of the protective measures listed apply to both direct and indirect contact, and others apply to one of these only.

Protection by insulation   This is the most usual means of providing protection against direct contact and is employed in most installations. Cables, electrical appliances, and factory-built equipment to recognised standards will normally comply with the requirements but it should be noted that paint or varnish applied to live parts will not provide adequate insulation for this purpose.

Protection by SELV-extra low voltage   Another means of protection is by separated extra low voltage which entails the use of a double wound transformer to BS EN 61558, the secondary winding being isolated from earth, and the voltage not to exceed 50 V.

The requirements of the IEE Regulation regarding SELV are modified in respect to equipment installed in bath and shower rooms. The arrangements for provision of switches and socket outlets are relaxed provided SELV is used at a nominal voltage not exceeding 12 V, and provided certain other conditions are met (IEE Section 601).

In certain circumstances protection may be by extra low voltage systems with one point of the circuit earthed. These are referred to as PELV and conditions for their application are contained in IEE Regulation 471-14.

Protection by obstacles or placing out of reach   Protection against shock can sometimes be achieved by the provision of obstacles which prevent unintentional approach or contact with live parts. These may be mesh guards, railings, etc. Another method of protection is to place live parts out of arms reach. This is defined in diagrammatic form in Part 2 of the IEE Regulations. These two methods may only be applied in industrial-type situations in areas which are accessible only to skilled or instructed persons. An example would be the exposed conductors for supply to overhead travelling cranes.

Protection by the use of Class 2 equipment   This is equipment having double or reinforced insulation, such as many types of vacuum cleaner, radio or TV sets, electric shavers, power tools and other factory built equipment made with 'total insulation' as specified in BS EN 60439.

Conductive parts inside such equipment shall not be connected to a protective conductor and when supplied through a socket and plug, only a two-core flexible cord is needed. Where 2-pin and earth sockets are in use it is important to ensure that no flexible conductor is connected to the earth pin in the plug. It is necessary to ensure that no changes take place which would reduce the effectiveness of the Class 2 insulation, since this would infringe the BS requirements and it could not be guaranteed that the device fully complies with Class 1 standards.

Protection by earthed equipotential bonding and automatic disconnection of supply Sometimes referred to as 'eebad', this method of protection is of general application, and is in fact the most commonly used method of protection against indirect contact. The object is to provide an area in which dangerous voltages are prevented by bonding all exposed and extraneous conductive parts. This is covered in IEE Regulation 413-02. In the event of an earth fault occurring outside the installation, a person in the zone concerned is protected by the exposed and extraneous conductive parts in it being electrically bonded together and so having a common potential. The same is not true when a fault occurs within the installation.

The practicalities of the bonding requirements, calculation of the sizes of bonding and protective conductors, and their installation is dealt with in Chapter 15 of this book.

For the protection to be effective it is necessary to ensure that automatic disconnection takes place quickly. This aspect is covered in IEE Regulation 471-08. For TN systems, disconnection times between 0.1 and 0.8 seconds are specified in Regulation 413-02-08 (depending upon the nominal voltage to earth) for hand-held or movable equipment. The disconnection may be extended to 5 seconds for distribution circuits and final circuits supplying certain types of equipment. IEE Regulation 413-02 gives detailed information on the various requirements, and IEE Tables 41B1, 41B2 and 41D give values of earth fault loop impedance for the different conditions and types of protection used. An alternative means of providing protection against indirect contact is given in IEE Regulation 413-02-12 with related values of circuit protective conductor impedence specified in IEE Table 41C. Compliance will ensure that with an earth fault the voltage appearing on the exposed conductive part does not exceed 50 V. Automatic disconnection is generally brought about by use of the overload protection device. To achieve a sufficiently rapid disconnection the impedance of the earth loop must be low enough to give the disconnection time required. An alternative way of doing this is by the use of a Residual Current Device.

Residual current circuit breaker As stated above, rapid disconnection for protection against shock by indirect contact can be achieved by the use of a Residual Current Device. A common form of

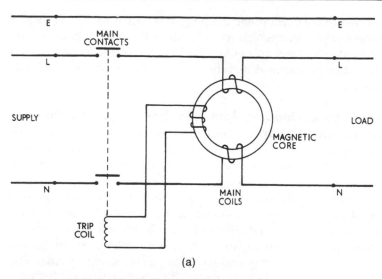

(a)

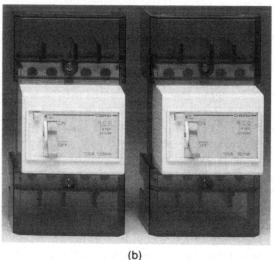

(b)

Fig. 2.6 *(a) A simplified diagram of a residual current circuit breaker (RCCB); (b) illustrations of RCCBs with 100 mA and 30 mA residual current ratings (Crabtree Electrical Industries Ltd)*

such a device is a residual current circuit breaker and a simplified diagram of a typical device is shown in Fig. 2.6. The method of operation is as follows. The currents in both the phase and neutral conductors are passed through the residual current circuit breaker, and in normal operating circumstances the values of these currents are equal. Because the currents balance, there is no induced current in the trip coil of the device. If an earth fault occurs in the circuit, the phase and neutral currents no longer balance and the residual current which results will cause the operation of the trip coil of the device. This will in turn disconnect the circuit by opening the main contacts.

IEE Regulations call for Residual Current Devices to be used to protect any socket which can be expected to be used for supplying outdoor equipment (IEE Regulation 471-16) and is preferred for any socket outlets which are part of a TT system (IEE Regulation 471-08-06). Residual current devices may also be used if difficulties are experienced in obtaining sufficiently low earth fault loop impedance to obtain a satisfactory disconnection time.

Residual current circuit breakers used to be known as current operated earth leakage circuit breakers. This term is now obsolete but equipment so marked may still be encountered. It should be noted that earth leakage circuit breakers of the *voltage* operated type are no longer permitted under the IEE Regulations, and must not be fitted. Voltage operated devices which may have been fitted in the past may need testing and advice on this is given in Chapter 16.

It should be noted that Residual Current Devices cannot be used where a PEN (combined protective and neutral) conductor is in use on the load side of the RCD for the simple reason that even in earth fault conditions the currents will balance and there will be no residual current to operate the breaker (IEE Regulation 413-02-07).

Protection by non-conducting locations   This method is limited in application to such situations as electrical test rooms which are under effective supervision. In the non-conducting location there shall be no protective conductor and all exposed conductive parts shall be permanently arranged so that persons will not come into simultaneous contact with two such parts or between these and any extraneous conductive parts. Two metres is the minimum distance required between the parts mentioned above. All floors and walls must be of insulating materials.

Protection against thermal effects

Chapters

42 & 55

Items of fixed equipment shall be erected so as not to inhibit its intended heat dissipation.

Luminaires and lamps shall be adequately ventilated, and spaced from wood or other combustible materials.

Fixed equipment containing flammable dielectric liquids exceeding 25 litres should have provision for safely draining any spilt or surplus liquid, and should be placed in a chamber of fire resisting construction if within a building.

The IEE Regulations include a table which gives temperature limits for accessible parts of equipment. These range from 55 to 90°C, dependent on whether the equipment is likely to be touched, and any equipment which will exceed the limits must be guarded so as to prevent accidental contact:

Protection against overcurrent

Chapter   Sections

| 43 | — | 473 & 533 |

Every electrical circuit and final circuit shall be protected against overcurrent by suitable overcurrent protective devices (IEE Regulation 130-04-01). These devices could be circuit breakers to BS EN 60947-2: 1996, HBC fuses to BS 88 or BS 1361, or rewirable fuses to BS 3036 (Fig. 2.7).

Overcurrent may be divided into two distinct categories, overload and short circuit. Overload current is an overcurrent occurring in a circuit which is electrically sound. For example the current caused by an electric motor which is stalled. Short circuit current is that which arises due to a fault in the circuit, as with a conductor which has become disconnected, or in some other way shorted to another, causing a very low resistance fault between the conductors. The IEE Regulations deal with overcurrent in Chapter 43 and overload and short circuit are in Sections 433 and 434 respectively. When considering circuit design both aspects of overcurrent have to be taken into account.

It is often possible to use the same device to protect against overload and short circuit, but before doing so it is necessary to determine the design current of the circuit and also to ascertain the prospective short circuit current which is likely to arise.

Overload protection   Overload protection is intended to prevent the cables and conductors in a circuit from undue temperature rise, and it is necessary to ensure that rating of the device chosen is suitable for this. Having determined the normal current to be drawn by a circuit, the cable installed must be able to carry at least that value.

Fig. 2.7   *Overcurrent protective devices. Miniature circuit breaker (left), a BS 1361 fuse (second left) and rewirable fuses to BS 3036 (right). The use of HBC fuses or miniature circuit breakers is strongly recommended. At centre is shown a residual current circuit breaker (George H. Scholes PLC)*

The protective device in its turn must be able to protect the cable chosen. For example, a circuit may be expected to carry a maximum of 26 A. The cable chosen for the circuit must be one which will carry a larger current, say, 36 A. The overload device must be rated at a figure between the two so that it will trip to protect the cable but will not operate under normal conditions. In the case quoted a 32 A MCB or HBC fuse would be suitable. A device provided for overload protection may be installed at the start of the circuit or alternatively near the device to be protected. The latter is common in the case of electric motors where the overload protection is often incorporated in the motor starter.

In some special circumstances it is permissible to omit overload protection altogether, and IEE Regulations 473-01-04 and -03 cover this. In some cases an overload warning device may be necessary. An example given is the circuit supplying a crane magnet where sudden opening of the circuit would cause the load on the magnet to be dropped.

In a few cases protection is afforded by the characteristics of the supply. Supplies for electric welding come into this category, where the current is limited by the supply arrangements and suitable cables are provided.

Short circuit protection   Short circuit currents need to be broken before thermal or mechanical damage can be caused to conductors or connections. The value of a short circuit current is potentially very high, usually of several thousand amperes, due to the resistance of the circuit being very low, comprising as it does only the cables feeding the fault. Protective devices, whether they are fuses or circuit breakers, all have a rated breaking capacity. This is the maximum fault current which the device is capable of clearing. It is important to ensure that under fault conditions the prospective current resulting from a dead short does not exceed the breaking capacity of the device. This is stated in IEE Regulation 434-03, although there is an exception which we shall come to in a moment.

The protective device for short circuit protection must be placed close to the origin of the circuit concerned (IEE Regulation 473-02). To ensure that a suitable protective device is used it is necessary to ascertain the maximum level of short circuit current which may occur. There are a number of ways of doing this, and it would be normal to consult the supply authority to ascertain the value of the prospective short circuit current at the origin of the installation. Alternatively, it is possible to calculate a value knowing the supply voltage and the resistance of the conductors in the circuit. It will be difficult to ascertain the resistance of the supply authority's cables back to the supply transformer, but an approximation can be made by considering only the supply cables which are within the boundary of the property concerned. By assuming the resistance of the remaining parts of the supply network is zero, the value of the short circuit current derived will be on the safe side when used to select the protective device.

The short circuit current can be calculated using ohms law thus:

$$\text{Prospective short circuit current} = \frac{\text{supply voltage}}{\text{circuit resistance}}$$

For example, if the total resistance of the cables were to be 0.02 Ω, and the supply voltage 230 V the short circuit current would be 11 500 A. A similar method can be used to calculate the prospective short circuit current at a point other than at the origin, and in this case the resistance of the additional cables is also used in the calculation. Typical breaking capacities for protective devices are 80 000 A for HBC fuses to BS 88, up to 16 000 A for miniature circuit breakers, and between 1000 to 4000 A for rewirable fuses. The manufacturer will be able to provide details for particular devices.

An important IEE Regulation is number 434-03-02 which relates to overload protection also being used as short circuit protection. Provided the requirements for overload are satisfied and the device has an adequate breaking capacity, it can be assumed that suitable short circuit protection has also been provided. In the cases when it is not possible to comply with Regulation 434-03-02, it will be necessary to check that the temperature rise of the conductors does not become excessive due to a short circuit fault. This is covered in IEE Regulation 434-03-03, which gives the necessary formula and data.

In installation practice, fuses or other protective devices are often connected in series, and in this case the breaking capacity of the downstream device is allowed to be less than the prospective short circuit current provided certain conditions are met. This is permitted because under short circuit conditions, the protection is designed to operate very rapidly, and, depending upon the type of protective device used, the circuit is broken before the full fault current builds up. The energy passed in the short space of time prior to the circuit being fully broken is termed the 'let-through' energy, and advantage can be taken of this fact provided the ratings of the two devices are coordinated.

Data on the breaking capacity and let-through energy of the various protective devices which are available, can be obtained from the manufacturers concerned.

Rewirable or semi-enclosed fuses   Table 3.1 gives the sizes for fuse elements of copper wire used in semi-enclosed fuses. These ratings are based upon the maximum current in the circuit, and not the current at which the fuse will operate. It will be seen for example, that a 15 A circuit will be protected by a fuselink consisting of 0.50 mm copper wire. This wire is designed to carry 15 A without overheating, and will not melt until 29 A flows through the circuit. If a fuse element which melted at 15 A were provided in this circuit it would operate as soon as the normal current in the circuit was reached, and even before the circuit becomes fully loaded the wire might become red hot and cause considerable damage to the switchgear.

When rewirable fuses are employed it is a well known fact that the most common cause of overheating and breakdown of switchgear is due to the fitting of incorrect sizes of fusewire. The IEE Regulations do not exclude the use of semi-enclosed fuses, but PVC insulated and other types of cables in certain circumstances need to be derated as will be seen on referring to the IEE Tables of current ratings of cables.

The only advantage of the rewirable fuse is that it is cheap, and costs practically nothing to renew, but these savings are generally offset against the costs of the large cables which may be necessary when the fuse is used for overload protection. There is nothing else that can be said in favour of this type of protection.

Another disadvantage is that it has a low rupturing capacity and cannot be relied upon to clear heavy faults.

Supply authorities no longer rely upon this type of fuse for protection, and the time is not far distant when this rather crude method of overcurrent and short-circuit protection will be a thing of the past.

HBC fuses   HBC fuses to BS 88 and cartridge fuses to BS 1361 (Fig. 2.8) will give discriminate protection against overcurrents, and will also clear short-circuit currents rapidly and safely up to their rated breaking capacity.

Special HBC fuses are sometimes needed for motor circuits to take care of heavy starting currents, and normal overcurrent protection

Fig. 2.8   *A range of small sizes of HBC fuses. HBC fuses to BS 1361 together with a holder are shown at top, and HBC fuses to BS 88 are illustrated below*

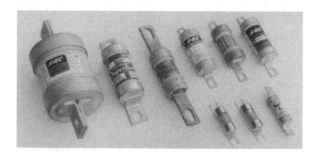

Fig. 2.9   *A range of HBC fuses to BS 88 including the larger sizes used in industrial installations. They are used on low-voltage circuits where high short circuit currents are possible, and as back-up devices for circuit breakers, for protecting electric motors, and for many similar applications*

for these circuits is provided in the motor starters (see Table 5.3 and 'Final circuits feeding motors' in Chapter 5).

Miniature circuit breakers   Circuit breakers have characteristics similar to HBC fuses, and they give both over-current protection and short-circuit protection. They are normally fitted with a thermal device for overcurrent protection, and a magnetic device for speedy short-circuit protection.

A typical time/current characteristic curve for a 20 A miniature circuit breaker (MCB) is shown in Fig. 2.10 together with the characteristic for a 20 A HBC fuse. The lines indicate the disconnection times for the devices when subject to various fault currents.

The components of a typical MCB are illustrated in Fig. 2.11, and the method of operation is described in the caption. MCBs are available with breaking capacities up to 16 000 A, but the manufacturers' data should be consulted to determine the rating for a particular device. MCBs can be obtained combined with residual current devices, and these can be useful where RCD protection is a requirement.

Protection against undervoltage

Chapter   Sections

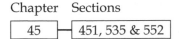

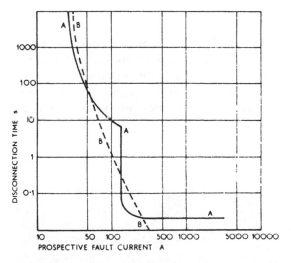

Fig. 2.10   *Typical characteristics of a miniature circuit breaker (line A) and an HBC fuse (line B). Both are for 20 A rated devices*

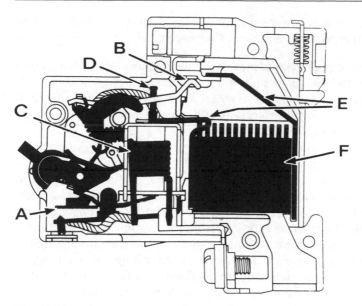

Fig. 2.11 *Cut-away section of a typical miniature circuit breaker. Under overload conditions a bi-metal blade operates the trip lever (A) which releases the contacts (B). Under short circuit conditions, the solenoid (C) is energised which opens the contacts (B) rapidly by means of plunger (D). The arc which occurs between the parting contacts is moved magnetically along the arc runners (E) and is split up and extinguished by the arc splitter pack (F) (MEM Ltd)*

An assessment needs to be made of the likelihood of danger arising from a drop in voltage, or loss and subsequent reinstatement of supply, and this may need to be done in conjunction with the user of the installation. Suitable protection may be provided by the use of 'no-volt' or 'low-volt' relays and the IEE Regulations lay down certain conditions for their operation.

Isolation and switching

Chapter	Sections
46	476 & 537

Suitable means of isolation must be provided so that the supply may be cut off from every installation (IEE Regs 131-14 and 460). Also, for every electric motor, efficient means of disconnection shall be provided which shall be readily accessible. Chapter 46 of the IEE Regulations deals with the subject in four categories, these being Isolation, Switching for mechanical maintenance, Emergency switching and Functional switching. In many cases one device will be able to satisfy more than one of the requirements. Switching for operating convenience, sometimes termed 'functional switching' or 'control switching' is covered in IEE Regulation 464 and in certain cases it may be possible to use functional switches as isolating devices provided they comply with the Regulations.

A distinction exists between Isolation, and Switching for mechanical maintenance. The former is intended for operation by skilled persons who require the circuit isolated so as to perform work on parts which would otherwise be live, whereas the latter is for use by persons who require the equipment disconnected for other reasons which do not involve electrical work. In both cases it may be necessary to provide lockable switches or some other means of ensuring the circuit is not inadvertently re-energised, but there are some differences in the types of switch which may be used.

Emergency switching is required where hazards such as rotating machinery, escalators, or conveyors are in use. Suitable marking of the emergency switch is required, and it is often the practice to provide stop buttons in suitable positions which control a contactor.

Every installation must be provided with means of isolation but the requirement for the other two categories is dependent upon the nature of the equipment in use. In all but small installations it will be necessary to provide more than one isolator so that the inconvenience in shutting down the whole installation is avoided when work is required on one part of it. If insufficient isolators are provided or if they are inconveniently placed there may be a temptation to work on equipment whilst it is still live.

Table 2.1 indicates the switching and other devices which may be used for the various functions. It should be noted that some types of circuit breaker, microbreak switches and semiconductor devices may not be used as isolating devices.

Circuit design

Having outlined some of the main requirements of the IEE Regulations for Electrical Installations it is now proposed to go into the more practical aspects of installation design. The main considerations are to determine the correct capacity of switchgear, protective

Table 2.1   *Switching and other devices permissible for the various purposes listed in the IEE Regulations*

Device	Use as isolation	Switching for mechanical maintenance	Emergency switching
Isolator	yes	–	–
Isolating switch	yes	yes	yes
Switch	–	yes	yes
Plug and socket	yes	yes	–
Fuse link	yes	–	–
Link	yes	–	–
Circuit breaker	see note A	yes	yes
Remote control of contactor	–	see note B	yes

Note A: Only if the circuit breaker is designed with the required contact separation (IEE Regulation 537-02-02).
Note B: It can be dangerous to use control circuits for isolation purposes or for mechanical maintenance unless additional precautions are taken.

devices and cable sizes for all circuits. In order to do this it is necessary to take into account the following:

Subdivision and number of circuits
Designed circuit current
Nominal current of protective device
Size of live conductors
Overload protection
Grouping factor
Ambient temperature
Thermal insulation
Voltage drop
Short circuit protection
Earth fault protection
Protection against indirect contact.

Subdivision and number of circuits

Section

314

Even the smallest of installations will need to be divided into a number of circuits. This is necessary for two reasons, firstly the need to

Fig. 2.12 *Modern all-electric kitchens often require a separate ring circuit to cater for the expected loads. Dish washers, clothes washers and dryers are often encountered in addition to the equipment shown here (Thermoplan (UK) Ltd)*

divide the load so that it can be conveniently and safely handled by the cabling and switchgear, and secondly the need to achieve the IEE Regulation requirement to take into account likely inconvenience (IEE Regulation 314-01-02).

In small installations it is appropriate to provide a minimum of two lighting circuits, so that in the event of a protective device tripping under fault conditions, a total blackout is avoided. In addition separate circuits must be provided for lighting and power. In the past there was a tendency to provide insufficient socket outlets, with the result that a proliferation of adaptors and flexible cords was used by the consumer. This led to a variety of hazards from overloading of sockets to tripping over the leads. This situation is undesirable and can only be avoided by installing sufficient outlets for the anticipated use. Also it should be noted that with the continuing increase in the use of electrical equipment in the home, loads are tending to rise, and this is particularly so in kitchens. It is recommended that a separate ring circuit is installed for the kitchen area to handle the loads arising from electric kettles, dish washers, washing machine and other equipment which is likely to be used. In deciding the number of arrangements of ring circuits to be provided, the designer must in any case consider the loads likely to arise so that an appropriate number of circuits is installed.

In larger commercial or industrial installations, it is also necessary for the division of circuits to be considered carefully. Following the assessment of general characteristics described earlier, a knowledge of the use of the installation and the nature of the processes which are to be undertaken will guide the designer in the choice of circuit division. In many commercial or industrial installations main and submain cables will be needed to supply distribution boards feeding the range of final circuits.

Multicore cables in parallel

It is sometimes desirable to connect two or more multicore cables in parallel. Potential advantages are in cost or ease of installation as the cables used will be smaller than a single cable for the same duty and will have smaller bending radii. However the cables do have to be derated for grouping. Consider a circuit protected by a 500 A Circuit breaker. This would require a 300 mm 4-core PVC SWA and PVC cable. If parallel cables were to be installed, two smaller cables could be used. The calculation is shown below. In this example the derating factor is 0.86 and it is assumed the cables are to be clipped direct to a cable tray.

$$I_t = \frac{500\,\text{A}}{0.86 \times \text{cables}} = 291\,\text{A}$$

where I_t is the calculated current-carrying capacity required for each cable.

An inspection of IEE Table 4D4A shows that two 150 mm^2 cables will be required since a 120 mm^2 cable will only carry 290 A. If it is possible to install the two cables on the tray so that there is at least one cable diameter between the cables throughout their length – including terminations – the derating factor becomes 0.91 and two 120 mm^2 cables can be used. These cables could cost less than the original 300 mm^2 cable, and a further advantage is that the bending radius of the 120 mm^2 and 150 mm^2 cable is much less than the bending radius of the 300 mm^2 cable. Where there are confined spaces for the installations of large cables the reduced bending radius may well prove to be of a very considerable advantage.

There is no reason why two or more cables should not be connected in parallel but it is important to remember that IEE Regulations require that measures are taken to ensure that the load current is shared equally between them. This can be achieved by using conductors of the same material, cross-sectional area and length. Additionally, there

Fig. 2.13 *The use of cables connected in parallel ease the installation task and can often result in a more economical scheme. This view shows a distribution circuit employing parallel multicore 415 V cables, run on cable ladder, feeding a distribution board. The cables are 120 mm² multicore PVC insulated SWA copper. Had parallel cables not been used, 300 mm² cable would have been needed (W.T. Parker Ltd)*

must be no branch circuits throughout their length. It must also be remembered that socket-outlet ring circuits are not conductors in parallel. The larger the size of the conductor, the less current is carried per square millimetre of cross-sectional area.

For instance a 3-core armoured 95 mm² cable carries 2.64 A per mm² whereas a 400 mm² 3-core cable is rated to carry only 1.48 A per mm², therefore nearly half of the conductor of the larger cable performs no useful purpose.

In designing circuits with parallel cables, it is necessary to consider the effect of a fault condition in one conductor only. IEE Regulation 434-03-03 must be applied to check that the characteristic of the protective device is such that the temperature rise of the conductors under fault conditions is contained.

4-core cables with reduced neutrals

For multicore cables feeding polyphase circuits it is permitted to use a reduced neutral conductor, providing that there is no serious unbalance between phases, or providing these are not feeding computing equipment or discharge lighting circuits where significant harmonic currents are likely to occur (IEE Regulation 524-02-02).

Reduced neutrals may also be used when single-core cables are installed in conduit or trunking on 3-phase circuits.

When feeding panels which control 3-phase motors it is very often satisfactory to install a 3-core cable, and should 230 V be required at the panel for control circuits this can be provided via a small transformer. This is much cheaper than providing an additional core in a heavy cable but limits the future use of the cable.

Designed circuit current

Section

| 311 |

Having decided on the configuration and number of circuits to be used in the installation, the designer next needs to determine the design current for each circuit. Sometimes this can be quite straightforward, and in the case of circuits feeding fixed equipment such as water heaters, no particular problems arise. With circuits feeding socket outlets, the design current will be the same as the rating of the protective device since the designer has no control over the load a user may place on the socket outlet circuits.

One of the types of installation most difficult to assess is the modern industrial unit, commonly found nowadays on factory estates. In the case where the unit is new or unoccupied, there may be difficulty in determining its future use. It will then be necessary for the electrical designer to install capacity which is judged to be appropriate, informing the person ordering the work of the electrical arrangements provided.

Diversity has already been described on page 19 of this book. In assessing the current demand of circuits, allowance for diversity is permitted. It must be emphasised that care should be taken in its use in any particular installation. The designer needs to consider all the information available about the use of the building, processes to be used, etc., and will often be required to apply a degree of experience in arriving at the diversity figure which is appropriate. An example of the application of diversity is given in Chapter 4.

Another aspect which requires consideration is the allowance to be made for future growth. This may well be applied to the submain cable feeding a distribution board rather than a particular final circuit.

If it is probable that additional load will be added in the foreseeable future, allow for this by adding say, 25% to the maximum demand, although less than this may be appropriate in some cases.

To summarise, the following steps are required to determine the designed circuit current:

(1) Calculate the total installed load to be connected
(2) Calculate the assumed maximum demand after assessing diversity in the light of experience and information obtained
(3) Add an appropriate allowance for future growth.

Nominal current of protective device

Sections

432, 433 & 473

The nominal current of the protective device is based upon the designed circuit current as calculated above. As allowance has already been made for possible additional future load, the rating of the protective device (i.e. fuse or circuit breaker) should be chosen close to this value. IEE Regulation 433-02 demands that the nominal current of the protective device is not less than the design current of the circuit, and choosing a rating as close as possible to it should be the aim. A larger fuse or circuit breaker may not cost any more, but if one of a higher rating is installed it may be necessary to install a larger cable if the protective device is providing overload protection.

Size of live conductors

Sections   Appendix

433, 473 & 523 |—| 4

Some of the tables which are used to determine the cable size required are contained in IEE Appendix 4. The appendix is divided into sections. Tables in section 4A categorise the methods of installation of cables, the Tables 4B give correction factors for grouping of cables and the Tables 4C ambient temperature correction factors, and Tables 4D to 4L give the current ratings and associated voltage drops for a range of different cable types.

Table 4A1 categorises wiring systems into 20 different types. The current carrying capacity of the cables differs considerably according to the method of installation which is to be used. This needs to be borne in mind so that the appropriate columns are used when referring to Tables 4D and 4L. Extracts from IEE Tables 4B1, 4D2, and 4D4 are reproduced in Chapter 3 of this book.

Correction factors for ambient temperature are also given, these being shown in Table 4C2 if semi-enclosed fuses are used, and in Table 4C1 for the other types of protection. The size of cable to be used for any particular circuit will depend upon a number of factors, each of which is covered in the sections which follow. A selection of worked examples illustrating the various points is given later in the chapter.

Overload protection   Where the protective device is providing overload protection, the conductors of the circuit are sized to the nominal rating (I_n) of the protective device. However, it is a requirement of the IEE Wiring Regulations that the current required to operate the protective device I_2 must not exceed 1.45 times the current-carrying capacity of the circuit conductors I_z. A correction factor has therefore to be applied when the protective device does not disconnect the circuit within 1.45 times its rating. This correction factor is determined as follows:

Using the semi-enclosed (rewirable) fuse as an example,

$$\text{Fusing factor} = \frac{\text{Current required to operate the device } (I_2)}{\text{Nominal rating of the protective device } (I_n)}$$

For a rewirable fuse the fusing factor is 2.

$$\text{therefore } 2 = \frac{I_2}{I_n} \quad \text{therefore } 2I_n = I_2$$

Substituting for I_2 in the overload formula from IEE Regulation 433-02-02 as stated above:

$$I_2 \leqslant 1.45I_z \quad \text{becomes } 2I_n \leqslant 1.45I_z \tag{1}$$

where I_z is the tabulated current-carrying capacity of the conductor being protected. From (1) $I_n = 0.725 \, I_z$ which is more conveniently expressed in the form:

$$I_z = \frac{I_n}{0.725}$$

There is no symbol in the IEE Wiring Regulations for the 0.725 since it is not considered to be a derating factor, However, it is more convenient to consider it as a derating factor and the symbol $\underline{S}$ is a convenient way of remembering the $\underline{S}$emi-enclosed fuse factor. This means that where semi-enclosed fuses are used for overload protection – and only when they are used for overload protection – the derating factor 0.725 is applied as a divisor to the rating of the fuse (I_n) to give the current-carrying capacity required for the conductors (I_t). A similar calculation would be necessary for any other type of protective device which had a fusing factor larger than 1.45.

Where, due to the characteristics of the load, a conductor is not likely to carry an overload, overload protection does not need to be provided (IEE Regulation 473-01-04). This means that with a fixed resistive load the protective device is only required to provide protection against fault current. For example, in the case of an immersion heater the load is resistive and an overload is unlikely since it has a fixed load. Under these circumstances the protective device is only providing fault current protection. This means that the conductors can be sized to the design current of the circuit instead of the nominal rating of the protective device and the derating factor 0.725 for semi-enclosed fuses is not applicable. For example, if a 3 kW immersion heater is to be installed on a 230 V supply, what is the minimum current-carrying capacity required for the cable and what size of BS3036 semi-enclosed fuse will be required if no other derating factors apply?

$$\text{Current taken by immersion heater} = \frac{3000 \text{ W}}{230 \text{ V}} = 13 \text{A}$$

The fuse size required will be 15 A, but as it is only providing fault current protection since the load is a fixed resistive load, I_t for the cable is 13 A.

By inspection of the tables in IEE Appendix 4 it will be found that a 1.0 mm^2 cable will be suitable providing it is not in contact with thermal insulation.

Naturally the circuit must also satisfy the same limitations, placed on it by the IEE Wiring Regulations, as other circuits. As the fuse is only providing fault current protection a calculation will be required to check that the conductors are protected against fault current.

Where the protective device is providing overload protection then the derating factor 0.725 is applicable for semi-enclosed fuses. For example, if the design current of a circuit is 17 A and overload protection is provided by a 20 A BS 3036 semi-enclosed (rewirable) fuse then the minimum current-carrying capacity for the conductors would have to be:

$$I_t = \frac{20 \text{ A}}{0.725} = 27.59 \text{ A}$$

The MICC tables in Appendix 4 of the IEE Wiring Regulations are the only current-carrying capacity tables where the user is not instructed to use the factor 0.725 for the cables when they are being protected against overload. It therefore follows that the derating factor 0.725 is applied only to all other types of cable when a semi-enclosed fuse is providing overload protection.

If overload protection had been provided by an HBC fuse to BS 88 part 2 or 6 or a cartridge fuse to BS 1361 or an MCB, then the derating factor of 0.725 would not have been applied and the conductor could have been sized to the rating of the protective device, which in the above example was 20 A.

Having determined whether the circuit is protected against overload and fault currents or just against the latter, the other factors that affect the conductor size should be considered.

Grouping factor   Before circuits are checked for voltage drop, protection against fault currents and indirect contact, it is as well to first size the cables. If the cables are grouped with other cables they have to be derated. The amount they have to be derated is dependent upon how they are installed. Symbols used in the Wiring Regulations are now defined in IEE Part 2. The latest version for grouping is C_g although it seems sensible to just use the symbol G since this stands for grouping.

The derating or correction factor, as it is sometimes referred to, is used as a divisor and is divided into the rating of the protective device (I_n) or the design current (I_b) depending upon whether the circuit is protected against overload or not.

Where the cables are spaced twice the diameter of the larger cable apart, no derating for grouping is required or if the spacing between adjacent conductors is only one cable diameter the derating factor is reduced.

The derating factors for grouping will be found in IEE Tables 4B1, 4B2 and 4B3 in Appendix 4 and as an example of how to use these tables assume that three twin & CPC cables are to be bunched together and installed on the surface and have to be protected against overload by a 20 A MCB.

From IEE Table 4B1 the correction factor for three cables bunched together is 0.7. The calculated current-carrying capacity required for the cable (I_t) will therefore be:

$$I_t = \frac{I_n}{G} = \frac{20 \text{ A}}{0.7} = 28.57 \text{ A}$$

Using the same example let it be assumed that the cables are supplying fixed resistive loads and that the design current I_b (i.e. full load current) is 17 A what will the current-carrying capacity of the conductors have to be?

There is a difference between a circuit being protected against overload and a circuit that only requires short-circuit protection. In the latter case

$$I_t = \frac{I_b}{G} = \frac{17 \text{ A}}{0.7} = 24.29 \text{ A}$$

Thus when the circuit is only being protected against short-circuit current the current-carrying capacity calculated for the cable size is less, it being based on the circuit's design current instead of the protective device rating.

Ambient temperature   The tables giving the current-carrying capacity of cables in the IEE Wiring Regulations are based on an ambient temperature of 30°C, this being the ambient temperature used for the British Isles.

The ambient temperature is the temperature of the immediate surroundings of the equipment and cables before the temperature of the equipment or cables contribute to the temperature rise.

Two tables are provided in the IEE Wiring Regulations giving correction factors to be applied when the ambient temperature deviates from 30°C. IEE Table 4C2 gives the correction factors to be used when a BS 3036 semi-enclosed fuse is used as the overload protective device and IEE Table 4C1 gives the correction factors for all other types of protective device used for overload protection and for all devices (including BS 3036 fuses) when used only for short-circuit protection.

If the ambient temperature is going to be below 30°C, for example in a chiller room, the current-carrying capacity of the conductors can be increased and factors are given in the tables for ambient temperatures of 25°C.

The symbol used for the ambient temperature correction factor in the regulations is C_a, but again this is more easily remembered by the symbol A.

It may be necessary to measure the ambient temperature for instance, in roof voids or suspended floors where heating pipes are installed since the ambient temperature will be higher in such confined spaces. When measuring such temperatures it is essential that the measurements are not taken close to the items generating the heat, including any cables installed. If checking the ambient temperature for existing cables the measurement should be taken at least 0.5 metres from the cables in the horizontal plane and about 150 mm below the cables.

As before, the correction factor is applied as a divisor. For example, if a twin & CPC PVC cable is installed on a surface in a boiler room where the ambient temperature is 60°C and the circuit is provided with overload protection by a 20 A semi-enclosed fuse what current-carrying capacity must the cable have?

From IEE Table 4C2 for PVC cable at 60°C the factor is 0.69, therefore,

$$I_t = \frac{I_n}{A \times S} = \frac{20}{0.69 \times 0.725} = 39.98 \text{ A}$$

It should be noted that the semi-enclosed fuse factor S has to be included since it is at the origin of the circuit. Using the same example again but this time the fuse is only providing fault current protection.

In this case the ambient temperature correction factor is taken from IEE Table 4C1 and is 0.5 since the fuse is providing only short-circuit protection. The semi-enclosed factor S is not applicable for the same reason, therefore,

$$I_t = \frac{I_n}{A} = \frac{20}{0.5} = 40 \text{ A}$$

Thermal insulation   Where conductors are in contact with thermal insulation, the thermal insulation reduces the rate of flow of heat from the conductors, thus raising the conductor's temperature. This means that the current-carrying capacity of the conductor has to be reduced to compensate for the reduction in heat loss.

Where cables are to be installed in areas where thermal insulation is likely to be installed in the future, they should be installed in such a position that they will not come into contact with the insulation (IEE Regulation 523-04-01). If this is not practicable then the cross-sectional area of the conductor has to be increased. The amount the conductor is increased in size is dependent upon the manner in which it is in contact with the insulation.

Tables are given in Appendix 4 of the IEE Wiring Regulations for sheathed cables and single core cables enclosed in conduit where they are in contact with thermal insulation and on one side with a thermally conductive surface. The reference method given in the tables in IEE Table 4A1 is method 4.

Where cables are totally enclosed in thermal insulation they have to be derated, The Wiring Regulations give the symbol C_i for this, but it is easier to remember as T for Thermal insulation. Where the conductors are totally enclosed for 500 mm or more then the derating factor is 0.5. Where the cables are totally enclosed for distances less than 500 mm then the derating factor is obtained from Table 52A which gives the factors for 50 mm (0.89), 100 mm (0.81) 200 mm (0.68) and 400 mm (0.55), for values in between these it would be a matter of interpolation to give an approximate factor.

Example: A PVC twin & CPC cable is to be installed in a house to a 6 kW shower, the cable will be installed along the top of the timber joists in the ceiling clear of any insulation but will be totally enclosed in insulation for 100 mm where it passes through the ceiling to the shower unit. Calculate the current-carrying capacity required for the cable.

The connected load is 6000 W divided by 230 V = 26 A.

Since the shower is a fixed resistive load the protective device is only providing fault current protection so the cable is sized according to the design current I_b (full load current).

From IEE Table 52A the derating factor T for totally enclosed for a distance of 100 mm is 0.81 therefore,

$$I_t = \frac{I_b}{T} = \frac{26\text{ A}}{0.81} = 32.1\text{ A}$$

Multiple correction factors   So far each of the correction factors has been treated separately, but in practice several of the factors can affect the same circuit. The factors are, however, only applied to that portion of the circuit which they affect. Where each factor is affecting a different part of the circuit the designer can select the factor which will affect the circuit most and size the conductors by using that factor. In other situations a combination of the factors may affect the same portion of the circuit. It therefore follows that a common formula can be remembered that covers all the factors.

Using the simplified notation it will take the form:

$$I_t = \frac{I_n}{G \times A \times T \times S}$$ when overload and short circuit protection is

being provided and,

$$I_t = \frac{I_b}{G \times A \times T}$$ when only short circuit protection is provided.

Using the symbols G, A, T and S gives a useful mnemonic *GATS* which is easily remembered where:

G is the correction factor for grouping; Wiring Regs C_g
A is the correction factor for Ambient temperature; Wiring Regs C_a
T is the correction factor for thermal insulation; Wiring Regs C_i
S is the correction factor for semi-enclosed fuse; Wiring Regs – no
  symbol.

In practice the designer will try and avoid those areas where a derating factor is applicable, since the application of several factors at the same time would make the cable very large such that, in final circuits, they would not fit into accessory terminations.

There is one combination that has not yet been discussed and that concerns 13 A socket-outlet circuits. Standard circuit arrangements are given in Table 8A of the IEE *On-site guide*. First, the total load of a socket-outlet circuit is taken as the rating of the protective device protecting the circuit. Secondly the user of the socket-outlet circuit should be advised as to the maximum load that can be connected to the circuit so that he can ensure $I_b \leq I_n$.

Thirdly, irrespective of the size of the protective device, socket-outlet circuits can always be overloaded and diversity cannot be taken into account, as this has already been allowed for in the standard circuit arrangements. No derating of the circuit cables for grouping is required where two circuits are grouped together (four live conductors). The cables must be derated where the ambient temperature exceeds 30°C. Although not mentioned in the IEE *On-site guide* derating has to be carried out if the cables are installed in contact with thermal insulation.

Where more than two radial circuits are grouped together derating is worked out as for any other circuit protected against overload. Where more than two ring circuits are grouped together the calculation is different. As far as ring circuits are concerned the amount of current in each leg of the ring is dependent upon the socket-outlet distribution round the ring for instance, if all the load was at the mid-point of the ring then the current distribution would be 50% on each leg. As far as standard circuit arrangements are concerned the distribution is taken as being two thirds in one leg (0.67). The formula therefore for sizing the ring circuit conductors is:

$$I_t = \frac{I_n \times 0.67}{G \times A \times T}$$

No diversity in the circuit is allowed since it has already been taken into account. The circuit does not have to be derated when a semi-enclosed fuse is used as the protective device.

Relaxation to grouping factors

Now that all the types of derating factors have been considered, consideration can be given to some relaxations for cables or circuits grouped together. These relaxations are given in Appendix 4 of the Wiring Regulations and are illustrated below.

Grouped cables not subject to simultaneous overload

Where it can be guaranteed that not more than one circuit or cable in the group can be overloaded at any one time, i.e. not subject to simultaneous overload, then the following formulae can be used. Two calculations are required and the calculation that gives the largest I_t is the one used to select the cable size.

Calculation when the protective device is a fuse to BS 88 or BS 1361 or an MCB to BS EN 60898.

$$I_{t1} \geq \frac{I_b}{G \times A \times T} \tag{1}$$

$$I_{t2} \geq \frac{1}{A \times T} \sqrt{I_n^2 - 0.48I_b^2 \left(\frac{1-G^2}{G^2} \right)} \tag{2}$$

Whichever is the larger, I_{t1} or I_{t2}, is the one that is used to size the cable.

Calculation when the overload protective device is a semi-enclosed fuse.

Again where it can be guaranteed that not more than one circuit or cable can be overloaded at any one time, and the protective device is a semi-enclosed fuse providing overload protection the following formulae are used:

$$I_{t1} \geq \frac{I_b}{G \times A \times T \times S} \tag{3}$$

$$I_{t2} \geq \frac{1}{A \times T} \sqrt{1.9 I_n^2 - 0.48 I_b^2 \left(\frac{1 - G^2}{G^2} \right)} \tag{4}$$

Again the larger of I_{t1} or I_{t2} is chosen to size the cable.

Care is needed when deciding whether to use this formula to size cables. Where there are a number of socket-outlet circuits grouped together simultaneous overloading could occur since the designer has no control over what the user of the circuit plugs into it. For example, the use of the ring circuits may only be for electric typewriters, desktop computers, printers, etc., but if there is a failure of the central heating system then staff may plug in electric heaters to enable work to continue. In any event where protective devices are providing overload protection it is difficult to ensure that only one of the circuits will have an overload occurring at any one time.

Lightly loaded conductors   Another relaxation concerns lightly loaded conductors. In this instance if a cable will not carry more than 30% of its grouped rating, it can be ignored when counting the remaining cables in the group. This relaxation can be very useful where large numbers of control cables are installed in trunking or as armoured cables bunched with power cables.

Using the symbol I_{tab} for the current-carrying capacity of the cables as given in the tables in IEE Wiring Regulations Appendix 4, the following procedure can be used. Two calculations are required.

First the size of the lightly loaded cable has to be determined from the formula previously given, i.e. $I_t = I_n/G$ where G is the grouping factor for all the cables in the group and I_n is the protective device size for the lightly loaded cables. Having sized the lightly loaded cable the current-carrying capacity given in the table I_{tab} is noted.

A test is now made by a second calculation to see whether it will be carrying more than 30% of its grouped rating.

$$\text{Grouped current-carrying capacity } \% = \frac{100 \times I_b}{I_{tab} \times G}$$

If the result of this calculation is 30% or less the cable can be ignored when counting the rest of the cables grouped together. It is important to remember that the size of cable worked out in the first calculation for the lightly loaded cable must be used even if it appears to be too large for the current the cable is carrying.

Example: PVC insulated single core copper cables are used for sixteen single-phase control circuits fused at 2 A and are installed in trunking with four power circuits wired in PVC single core cable. If the design current for the control circuits is 1.8 A and the power circuits are protected by 32 A HRC fuses what size cables can be used?

First determine the minimum size of the control circuit cable.

Total number of circuits $= 16 + 4 = 20$

From Table 3.4 the correction factor for 20 circuits in trunking $= 0.38$. Calculated current-carrying capacity required for control circuits,

$$I_t = \frac{I_n}{G} = \frac{2 \text{ A}}{0.38} = 5.26 \text{ A}$$

From IEE Table 4D1A 11.0 mm^2 cable can be used with an I_{tab} of 13.5 A. Now test to see whether the current it is carrying is not more than 30% of its grouped rating.

$$\text{Grouped \%} = \frac{100 \times I_b}{I_{tab} \times G} = \frac{100 \times 1.8}{13.5 \times 0.38} = 35\%$$

This is more than 30% so the cables cannot be excluded from the group in which case the current-carrying capacity required for the power cables is:

$$I_t = \frac{32 \text{ A}}{0.38} = 84.2 \text{ A}.$$

From Table 3.4 cable size required for power cables is 25 mm^2. Try using 1.5 mm^2 cable for the control circuits (this is the minimum size industrial contractors would normally use). The first calculation does not need to be repeated. The I_{tab} for the 1.5 mm^2 cable is 17.5 A therefore

$$\text{Grouped \%} = \frac{100 \times I_b}{I_{tab} \times G} = \frac{100 \times 1.8}{17.5 \times 0.38} = 27\%$$

which means that if 1.5 mm^2 cables are used for control circuits they can be ignored when counting the number of other cables grouped together.

An alternative way of working out the second calculation once the lightly loaded conductor size has been determined is:

$$I_b < 0.3 \times G \times I_{tab} = 0.3 \times 0.38 \times 17.5 = 1.995 \text{ A}$$

and since I_b is only 1.8 A this also proves the 1.5 mm^2 cables can be ignored when determining the cross-sectional area of the power cables.

The derating factor for the power cables now becomes 0.65 for four circuits.

$$I_t = \frac{32 \text{ A}}{0.65} = 49.2 \text{ A}$$

This is a considerable reduction on the previous value of 84.2 A. The cable size required now is 10 mm^2.

Voltage drop

Section   Appendix

525	4

Having determined the size of conductors to install it is now necessary to check that the conductors chosen will comply with the voltage drop constraints demanded by the IEE Wiring Regulations before any further calculations are carried out.

IEE Section 525 covers voltage drop and Regulation 525-01-01 specifies that under normal operating conditions the voltage at the current-using equipment's terminals must not be less than that specified in the British Standard for that equipment. Where the input voltage is not specified for equipment by British Standards then the voltage at the equipment must be such as to ensure the safe functioning of the equipment. Where the supply is provided in accordance with the Electricity Safety, Quality & Continuity Regulations 2002, the voltage shall not vary by more than 10 percent above or six percent below the declared voltage. Regulation 525-01-01 may be considered satisfied if the voltage drop from the origin of the installation up to the fixed equipment does not exceed 4% of the declared nominal voltage. When calculating voltage drop, motor starting currents or in-rush currents to equipment can be ignored. However, the additional voltage drop caused by these currents has to be considered to ensure the satisfactory starting of equipment.

The Regulations do not mention diversity, but diversity can be taken into account when calculating voltage drop, since voltage drop is directly associated with the actual current flowing in conductors.

In general a 4% voltage drop allowance is used, but it is up to the designer to ensure that the voltage at the equipment terminals complies with Regulation 525-01-01 as stated above. Where the 4% value is used the single-phase voltage drop allowed is 9.2 V for 230 V supplies and the three-phase voltage drop 16 V for 400 V supplies.

All of the current-carrying capacity tables in Appendix 4 of the IEE Wiring Regulations have suffix A after the table number. A voltage drop table is also provided, this having a suffix B after the same table number. A sample current-carrying capacity table is given as Table 3.2. The number of this table in the IEE Wiring Regulations is 4D2A. A sample of the voltage drop table is also given, this being shown as Table 3.2 continued; the number in the IEE Wiring Regulations for this table is 4D2B.

On inspection of Table 3.2 it will be seen that the voltage drops are given in millivolts (mV) per amp per metre. These voltage drops are based on the conductor carrying its tabulated current I_{tab} so that the conductor's operating temperature is 70°C.

It will also be noticed that for cables up to 16 mm² only one value is given; this is the value using resistance only since it is considered that the reactance in such small cables has a negligible effect. For sizes larger than 16 mm² three values are given: resistance (r), reactance (x) and impedance (z). The voltage drops are also given in columns; column 2 for two core d.c. cables, column 3 for two-core single-phase a.c. and column 4 for three- or four-core three-phase a.c.

The formula for voltage drop is $V_d = L \times I_b \times mV/A/m \times 10^{-3}$

where L is the length of the circuit, I_b is the full load current and MV/A/m is from the appropriate table; the 10^{-3} converting millivolts into volts (i.e. division by 1000).

Two examples will suffice to illustrate how these voltage drop tables are used.

Example 1: A 230 V single-phase circuit using a 16 mm² twin & CPC cable with copper conductors, feeding a single-phase distribution board is 30 metres long and carries 80 amps. If the cable is clipped direct and no derating factors apply what will the voltage drop be? From Table 3.2 column 3 the voltage drop mV/A/m for 16 mm² cable is 2.8

$$V_d = \frac{30 \text{ m} \times 80 \text{ A} \times 2.8}{1000}$$
$$= 6.27 \text{ volts which is less than the allowed } 9.2 \text{ V}$$

Example 2: A three-phase 400 V motor is to be wired in PVC SWA & PVC armoured cable having copper conductors. The full load current of the motor is 41 A and the length of the circuit is 20 metres. If the three-phase voltage drop up to the distribution board is 6 V what size cable can be used to satisfy voltage drop?

If the normal procedure is adopted of picking a cable and working out the voltage drop several calculations may have to be made before a suitable cable is selected, however, a short cut can be used to size the cable by rearranging the formula.

$$\text{mV/A/m} = \frac{V_d \times 1000}{L \times I_b} = \frac{(16 - 6.0) \times 1000}{20 \times 41 \text{ A}} = 12.2 \text{ mV/A/m}$$

now look in the voltage drop Table 3.3 for a three-core cable whose mV/A/m is less than 12.2, a $4\,\text{mm}^2$ cable with a mV/A/m of 9.5 appears to be satisfactory. However, when using this method, the current-carrying capacity of the conductor must always be checked. In the above example although the $4\,\text{mm}^2$ conductor is satisfactory for voltage drop it will not carry the current and a $6\,\text{mm}^2$ (with an I_tab of 42 A) conductor must be used. This highlights the need for this extra check.

Actual voltage drop = $(20\,\text{m} \times 41\,\text{A} \times 6.4 \times 10^{-3})$ = 5.25 V

Add to this the voltage dropped up to the distribution board

Total voltage drop = 11.25 V, which is less than the 16 V allowed

Motor circuits   Where motor circuits or circuits with a power factor are concerned allowance can be made for the power factor of the circuit when calculating voltage drop and this is why the r and x values are given in the voltage drop tables. Similarly where the cables are not carrying their tabulated current-carrying capacity I_tab, allowance can be made for the reduction in cable resistance and the effect this has on voltage drop. The calculations get a little involved and are beyond the scope of this book.

Ring circuits   Calculation of the voltage drop in a ring circuit, or any other socket-outlet circuit, can only be approximate since the designer will have no control over what equipment the user plugs into each outlet. An average figure can be worked out based on all the load being at the mid-point of the ring, or one with the load being spread evenly round it. Initially the case with all the load at the mid-point is calculated, with half the load flowing in half the ring. If the voltage drop limits are satisfied, there is no need to go into the complication of working out the case with the load spread.

Short-circuit protection

Chapter   Sections

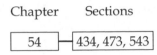

Having determined the size of the conductors the circuits require, the next stage is to check that they are protected against short-circuit current. It is then necessary to work out the size of protective conductors and then that the arrangements comply with the requirements for protection against indirect contact.

In the case of short-circuit currents the protective device is only providing protection against thermal and mechanical effects that occur due to a fault. It is therefore unnecessary for the conductors to be sized to the rating of the protective device. The relationship that the design current I_b must not be greater than I_n which in turn must not be greater than I_z is related only to overloads and has nothing to do with short-circuit protection.

Two calculations are required; one to give the maximum short-circuit current to enable switchgear or motor starters, etc., to be chosen with the correct fault rating and the second to determine the minimum short-circuit current in order to check that the conductors in the circuit are protected against short-circuit current.

Fortunately the calculation for checking that the conductors are protected does not have to be carried out in every circumstance due to IEE Regulation 434-03-02. Where the conductors have been sized to the overload requirements, which means that the conductor's current-carrying capacity is not less than the rating of the protective device, and that the breaking capacity of the protective device is not less than the prospective short-circuit current at the point the protective device is installed, it can be assumed that the conductors on the load side of the protective device are protected against short-circuit current, providing that if a circuit breaker is used, instead of a fuse, it is of the current-limiting type.

Obviously this rule will not apply to motor circuits where the protective device has to be sized to allow for starting currents, nor will it apply where, as in the example of the immersion heater, IEE Regulation 473-01-04 is used and the conductors are sized to the design current and not the protective device rating.

In these cases a calculation is required to check that the conductors are protected against short-circuit current. The minimum short-circuit current for single-phase and three-phase four-wire supplies will occur with a fault between phase and neutral and the formula is therefore:

$$I_{pn} = \frac{V_{ph}}{Z_{pn}}$$

where I_{pn} is the prospective short-circuit current between phase and neutral, Z_{pn} is the impedance of the phase and neutral from the source to the end of the circuit being checked and V_{ph} is the phase to neutral voltage.

Where the supply is three-phase three-wire then the minimum short-circuit current occurs between two phases and the formula now becomes:

$$I_{pp} = \frac{V_L}{2Z_p}$$

where I_{pp} is the phase to phase prospective short-circuit current, Z_p is the impedance of one phase only from the source to the end of the circuit being considered and V_L is the line voltage.

Checking that the conductors are protected involves working out how long it will take the protective device to disconnect the circuit and then comparing this time with the maximum time allowed for disconnection by the formula given in Regulation 434-03-03, i.e.

$$t \leq \frac{k^2 S^2}{I^2}$$

where t is the maximum disconnection time allowed, k is a factor dependent upon the type of conductor material, the initial temperature at the start of the fault and the limit temperature of the conductor's insulation, obtained from the same regulation. S is the cross-sectional area of the conductor and I the fault current.

Where the actual disconnection time is 0.1 secs or less then the actual energy let through the protective device has to be obtained from the manufacturer and this value must not exceed $k^2 S^2$.

Since it is the minimum short-circuit current that is needed, the resistance of the live conductors is taken at the temperature they will reach with the fault current flowing. This involves a complicated calculation so as a compromise the average is taken of the operating temperature of the conductor and the limit temperature of the conductor's insulation. For example, for standard PVC insulation with a copper conductor the maximum operating temperature allowed is 70°C, the limit temperature for the PVC insulation is 160°C so the average will be 115°C and this is the temperature used to determine the resistance of the conductors.

Earth fault protection The third form of overcurrent occurs when there is a phase to earth fault the rules for which will be found in Chapter 54 in the IEE Wiring Regulations. In this case there are two ways of checking that the thermal capacity of the protective conductor is satisfactory.

The first is to use Table 54G, an extract from which is reproduced in Table 15.2. If the protective conductor is made from a material which is different to the phase conductor, an additional calculation must be made. The size obtained from the table must be multiplied by k_1/k_2, where k_1 is the k factor for the phase conductor material and k_2 is the k factor for the material used for the protective conductor. However, in accordance with IEE Regulation 543-02-03 protective conductors up to and including 10 mm^2 have to be copper.

Apart from the 1.0 mm^2 size of cable none of the other PVC insulated and PVC sheathed cables containing a protective conductor (for instance, twin & CPC cable) comply with IEE Table 54G since the cross-sectional area of protective conductor is less than the phase conductor. The table can be used with aluminium strip armoured cables and for MICC cables since the cross-sectional area of the protective conductor at least complies with IEE Table 54G. For cables larger than 35 mm^2 the table can give a non-standard cable size, in which case the next larger size of cable is selected. In any event the table does not stop the phase earth loop impedance having to be worked out to enable protection against indirect contact to be checked.

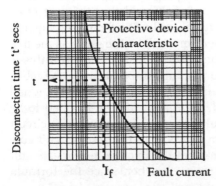

Fig. 2.14 *Finding disconnection time 't' using the curve showing characteristic of the protective device (Trevor Marks)*

The second method is to check that the protective conductor is protected against thermal effects by calculation by using the formula given in IEE Regulation 543-01-03.

$$S = \frac{\sqrt{I_f^2 t}}{k}$$

where S is the cross-sectional area required for the protective conductor, I_f is the earth fault current, t is the actual disconnection time of the protective device and k is a value from IEE Chapter 54 depending upon the conductor material, type of insulation, initial and final temperatures.

It is just a matter of comparing the actual cross-sectional area of the protective conductor with the minimum cross-sectional area required by calculation.

The value of I_f is determined from the formula $I_f = \quad I_f = \dfrac{V_{ph}}{Z_S}$

The value of Z_S is obtained from $Z_E + Z_{inst}$ where Z_E is the phase earth loop impedance external to the circuit being checked and Z_{inst} is the phase earth loop impedance for the actual circuit.

The disconnection time t is then obtained from the protective device characteristic as illustrated in Fig. 2.14. A line is drawn from the current axis for the value of I_f calculated, up to the characteristic for the protective device being used. At the point this line touches the characteristic it is then drawn horizontal to the time axis to give the actual disconnection time for that particular fault current.

The important thing to remember when choosing the value of k is the temperature of the conductor at the start of the fault. The protective conductor does not have to be carrying current to have its temperature raised. For instance, a twin & CPC cable has the protective conductor in the cable and although it will not be carrying current under normal conditions its temperature will be the same as the live conductors in the cable, and without additional information this is taken as the maximum permitted operating temperature for the cable.

Remember it is the minimum fault current that is important when proving conductors are protected. So if a calculation has had to be carried out to check that the circuit is protected against short-circuit current and the earth fault current is less than the short-circuit current, it will be necessary to recheck that the live conductors are still protected with the earth fault current. Further, since the formula used is the same as for short-circuit protection, if the disconnection time is 0.1 seconds or less then the manufacturer's let-through energy I^2t has to be compared with k^2S^2 in the same way as for short-circuit currents.

One further point that must be remembered is that it is important to take the impedance not the resistance of conduit, trunking and the armouring of cables.

Protection against indirect contact

Part   Chapter   Section

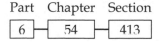

The basic requirement that has to be complied with is given in IEE Regulation 413-02-04 which states that the characteristics of the protective device, the earthing arrangements and the impedances of the circuit conductors have to be coordinated so that the magnitude and duration of the voltage appearing on simultaneously accessible exposed and extraneous conductive parts during an earth fault shall not cause danger.

This requirement falls into two separate subjects, the first is equipotential bonding and the second is automatic disconnection and it is this latter item that will now be dealt with first.

It must be clearly understood that equipotential bonding only allows the same voltage to appear on exposed and extraneous conductive parts within an installation if the fault is outside that installation. The voltage appearing on both the exposed and extraneous conductive parts is the fault current I_f times the impedance of the protective conductor from the source neutral up to the installation's main earth bar, to which the main equipotential bonding conductors are also connected. Where a fault occurs within the installation a voltage will appear on exposed conductive parts and this will be of a higher voltage than that appearing on extraneous conductive parts as illustrated in Fig. 2.15. From the figure it can be seen that the voltage appearing on the exposed conductive part is $I_f(R_2 + R_3)$, whereas the voltage appearing on the radiator is I_fR_3, the resultant potential difference is I_fR_2. This voltage difference may be as high as 150 V and is therefore quite dangerous.

The severity of the electric shock a person can receive is governed by three items: first the magnitude of the voltage, secondly the speed of disconnection and thirdly the environmental conditions.

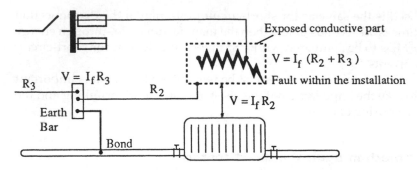

Fig. 2.15   *Shock voltage with bonding (Trevor Marks)*

To try and limit the severity of the electric shock when persons are in contact with exposed conductive parts whilst a fault occurs, the IEE Wiring Regulations specify maximum disconnection times depending upon the voltage to earth and the environmental conditions. The disconnection times for normal environmental conditions – such as those found in carpeted offices which are heated where the occupants are clothed and wearing socks and shoes – are given in IEE Regulation 413-02-08. Where the environmental conditions are not as described above then the disconnection times have to be decreased or other precautions taken. Such situations are deemed to be special situations and are covered in Part 6 of the IEE Wiring Regulations.

To determine that a circuit will disconnect within the specified time a calculation has to be carried out to determine the earth loop impedance at the end of the circuit. This involves knowing the phase earth loop impedance up to the origin of the circuit referred to as Z_E and the earth loop impedance of the circuit referred to as Z_{inst}. The overall earth loop impedance Z_S can then be determined as follows:

$$Z_s = Z_E + Z_{inst} \ \Omega$$

The value of Z_{inst} represents the impedance of the circuit's phase conductor Z_1 added to the impedance of the circuit's protective conductor Z_2. Tables are available giving $Z_1 + Z_2$ added together for different types of cables. As with earth fault currents it is important to take the impedance of conduit, trunking and the armouring of armoured cables. For circuits up to $35 \, mm^2$ impedances Z_1 and Z_2 above are replaced by resistances R_1 and R_2.

In practice there are two values of Z_S; there is the actual value as calculated or measured and there is the value given in the IEE Wiring

Regulations that must not be exceeded. The calculated value is compared with the value allowed to ensure the disconnection time is not exceeded.

In a simple installation, such as a house, the value of Z_E will be external to the origin of the installation. In a large commercial building or factory the value of Z_E will be the Z_S of the circuit feeding the bus bars of the section distribution board or the distribution board for the final circuit.

Resistance to be used in all calculations   Regulation 413-02-05 states that account has to be taken of the increase of resistance of the circuit conductors due to a fault current flowing through them. However, where a protective device complies with the characteristics in Appendix 3 of the Wiring Regulations and the earth loop impedance complies with the Z_S values given in the tables in Part 4 of the Regulations the circuit is deemed to comply with Regulation 413-02-05, i.e. the increase in resistance due to increase in temperature can be ignored. The difficulty is determining whether the protective device complies with the characteristics in Appendix 3 since the manufacturers' characteristics are not the same as those given in the Regulations. Additionally, the Z_S values in the tables in Part 4 of the Regulations will have to be adjusted if the open circuit voltage of the mains supply is different from the 240 V the tables are based upon. Where the characteristics do not comply with Appendix 3 of the Regulations or are not included in Appendix 3 then the resistance of the conductor is taken at the average of the operating temperature of the conductor and the limit temperature of the conductor's insulation. For example, for standard PVC insulation with a copper conductor the maximum operating temperature allowed under normal load conditions is 70°C, the limit temperature for the PVC insulation is 160°C so the average will be:

$$\frac{70 + 160}{2} = 115°C$$

Different types of insulation allow different maximum operating temperatures under normal load conditions, therefore the average has to be worked out for the type of insulation being used for example, the average for 90°C thermosetting insulation with a limit temperature of 250°C would be 170°C. Similarly the average will also need to be worked out for protective conductors and this will vary depending upon the type of protective conductor and how it is installed. Where a protective conductor is installed so that it is not in contact with other live conductors its operating temperature will be at ambient temperature which in Britain is taken as 30°C. The average in this instance will be 95°C. Where metal sheathed or armoured cables are installed the operating temperature of the sheath is taken as being 60°C, the limit temperature being dependent upon the type of conductor insulation. The tables in Chapter 54 of the IEE Wiring Regulations give the assumed initial temperature and limit temperatures for different conditions.

The average temperature worked out is the temperature used to determine the resistance of the conductors by using the formula:

$$R_{t_2} = R_{20}(1 + \alpha(t_2 - t_1)) \ \Omega$$

where t_2 is the final conductor temperature (i.e. 115), t_i is the temperature of the conductor's resistance R_{20} at 20°C and α is the resistance-temperature coefficient for both aluminium and copper from the simplified formula in BS 6360, which is 0.004. If the values $t_2 = 115$°C and $t_1 = 20$°C are put into the formula it will be found that the resistance R_{t_2} at 115°C is equal to the resistance R_{20} at 20°C multiplied by 1.38.

Reactance has to be taken into account on cables larger than 35 mm^2, but reactance is not affected by temperature so no adjustment for temperature has to be made to the reactance value.

Where large cables are used then the impedance of the cable has to be used, not just its resistance, the impedance being obtained from the formula:

$$Z = \sqrt{r^2 + x^2} \ \Omega$$

Summary

To recapitulate, the stages needed in determining the design of the distribution and final circuits are:

(1) Calculate the maximum demand taking into account diversity where appropriate.
(2) Determine from the electricity supply company whether a supply can be made available for the maximum demand required. Also obtain the following information:
   (a) The type of supply and frequency, is it single-phase or three-phase four-wire, 50/60 Hz.

(b) The earthing arrangement, i.e. the type of system of which the installation will be part.

(c) The rating and type of the cut-out (fuse) at the origin of the installation.

(d) The single-phase prospective short-circuit current at the origin (I_{pn}).

(e) The three-phase symmetrical short-circuit current if the supply is three-phase (I_p).

(f) The maximum earth loop impedance Z_E at the origin.

(3) Work out the distribution arrangement, trying to place distribution boards near the heaviest loads.

(4) Determine the type of protective devices that are going to be used throughout the installation.

(5) Determine which circuits are being protected against overload and short-circuit current and those which are being protected only against short-circuit current.

(6) Determine what derating factors are applicable to each circuit.

(7) Calculate the size of live conductors for each circuit.

(8) Calculate the voltage drop for each circuit, checking that it is acceptable.

(9) Check to ascertain that the conductors chosen are protected against short-circuit current.

(10) Calculate the size of protective conductors to be used throughout the installation.

(11) Check to ascertain that the circuits give protection against indirect contact.

(12) Size main equipotential bonding conductors and determine items to be bonded.

(13) Check to see if there are any special situations and if there are, size supplementary bonding conductors.

(14) Check to see whether any distribution boards are supplying circuits which are a mixture of stationary equipment, socket-outlets and/or portable equipment intended for manual movement during use, or hand-held Class I equipment and decide whether:

(a) to limit the CPC impedance in accordance with IEE Table 41C or,

(b) to install from the distribution board equipotential bonding conductors of the same size and to the same extraneous conductive parts as the main equipotential bonding conductors.

(15) Determine positions of switches, isolators and emergency stop buttons.

DISTRIBUTION BOARD REFERENCE:-					
1	**TYPE OF CONDUCTORS**				
	CIRCUIT DETAILS				
2	CIRCUIT REFERENCE				
3	LIGHTING LEVEL REQUIRED (LUX)				
4	REFLECTANCE LEVELS				
5	$R.I = \dfrac{ROOM\ L \times W}{H\ ABOVE\ WP \times (ROOM\ L + W)}$				
6	UTILISATION FACTOR				
7	MAINTENANCE FACTOR				
8	LAMP OUTPUT				
9	$NO\ OF\ LUMINAIRES = \dfrac{L \times W \times LUX}{LAMP\ OUTPUT \times MF \times UF}$				
10	NO OF SOCKET OUTLETS				
11	LOAD ALLOCATED PER SOCKET OUTLET				
	FACTORS				
12	TOTAL LOAD OF CIRCUIT $\left[3\ PHASE = \dfrac{kW}{\sqrt{3} \times V \times PF} \right]$				
13	LOAD WITH DIVERSITY				
14	FUSE / MCB TYPE				
15	FUSE / MCB RATING				
16	CORRECTION FACTORS				
	GROUPING				
	AMBIENT				
	THERMAL				
	DEPTH				
17	CONDUCTOR SIZE AND TYPE				
	VOLT DROP				
18	mV / A / M				
19	L.O.R.				
20	LOAD CURRENT (FROM 12 OR 13)				
21	TOTAL VOLT DROP				
22	TOTAL VOLT DROP (RING) 20×0.25				
23	MAXIMUM PERMISSABLE VOLT DROP				
	IMPEDANCE CONSTRAINTS				
24	MAX Zs (FROM 41B1, 41B2 ETC.)				
25	Ze (As given, measured)				
26	ACTUAL Zs				
27	CHECK $26 \leq 24$				
	THERMAL CONSTRAINTS				
28	FAULT CURRENT				
29	ACTUAL DISCONNECTION TIME				
30	MAXIMUM DISCONNECTION TIME (Regulations)				
31	k FACTOR				
32	$MINIMUM\ SIZE = \left[\dfrac{\sqrt{I^2 t}}{k} = mm^2 \right]$				
33	ACTUAL CABLE SIZE				
34	TOTAL FACTORS PHASE & CPC				
35	CONTAINER CAPACITY FACTOR				

Fig. 2.16   *Section of a typical design record sheet. The layout of the sheet enables the designer to systematically record assumptions and design data, and the calculation results*

This is a general list and there will be cases where all of the above items will not apply, for example, where IEE Regulation 473-02-04 allows certain circuits not to be protected against fault current. A typical design schedule is shown in Fig. 2.16.

Worked example

It is proposed to build a kitchen with an attached new boiler house at a factory. The assessment of the installation revealed the following information:

The nearest distribution board in the factory that is capable of taking the additional load is some 60 metres from the position at which the new distribution board will be placed in the kitchen. The supply to the existing board is 400 V three-phase four-wire, the single-phase short-circuit current is 3380 A and the three-phase short-circuit current is 5518 A. The phase earth loop impedance Z_S at the factory distribution board is 0.19 Ω and the three-phase voltage drop up to the board is 6 volts. The new feeder cable to the kitchen can be clipped direct to the surface of the structure and not grouped with other cables.

The proposed load on the kitchen is as follows:

Three-phase four-wire stationary equipment:

2–9 kW-hot plates, 1–11 kW-fryer, 1–12 kW-oven and 1–25 kW-dishwasher.

Single-phase stationary equipment:

1–1.5 kW-waste disposal unit, 1–3 kW-microwave oven, 2–1 kW-freezers, kitchen lighting 6 A.

Single-phase movable equipment and socket-outlets:

There will be a 32 A socket-outlet ring circuit to supply 4–1 kW-food mixers and 1–1 kW-slicer and sundry kettles, boiling pans, etc.

The boiler house equipment will be supplied from the kitchen distribution board the load comprising a 6 A single-phase supply to the boiler control panel, a 3 kW three-phase 400 V three-wire induced draught fan with a power factor of 0.64 and a 2.2 kW three-phase three-wire forced draught fan having a power factor of 0.61. The overloads in the starters for each motor being 4.5 A minimum, 6.75 A mid point and 9 A maximum.

The ring circuit is 38 metres long, all other circuits being 18 metres long. The installation from the kitchen distribution board being carried out in heavy gauge steel conduit.

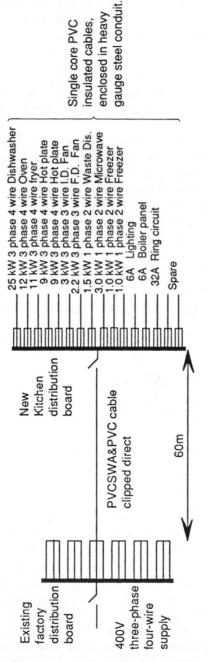

Fig. 2.17 *Distribution diagram for kitchen circuits referred to in the worked example*

Working:
The first requirement is to work out the maximum demand allowing for diversity where possible. The loads should also be balanced as far as possible over the three phases and since single-phase loads are also mixed in with the three-phase it pays to work it out in tabular form. The distribution arrangement will look like Fig. 2.17. For three-phase and neutral resistive load circuits the phase current is worked out – using the oven as an example – as follows:

$$I_L = \frac{\text{kW} \times 1000}{\sqrt{3} \times V_L} \text{ which for a 12 kW load} = \frac{12 \times 1000}{1.732 \times 415} = 16.69\,\text{A}$$

So a 20 A HRC fuse would be used.

Where motors are concerned the formula is slightly different since power factor has to be taken into account. Using the 3 kW induced draught fan as an example:

$$I_L = \frac{\text{kW} \times 1000}{\sqrt{3} \times V_L \times \text{p.f.}} = \frac{3 \times 1000}{1.732 \times 415 \times 0.64} = 6.5\,\text{A}$$

In this case the fuse has to take into consideration starting currents and 16 A fuses would be required. Similar calculations would be done for the other circuits. The load on each phase after diversity can then be listed to give the total load on each phase. In the table the revised load is the Actual load × % diversity. The results of the calculation are shown in Table 2.2.

Strictly speaking the addition in Table 2.2 should be a vectorial addition if the single-phase loads have a different phase angle to the three-phase loads, but in practice this fact is usually ignored.

This is probably as balanced a load as is possible, bearing in mind that although the ring circuit can take 32 A it will not be taking this amount all the time and there will be occasions when the sockets will not be taking any load at all.

At this point consideration should be given as to what future demand is likely to be required, in this instance allow say 10%. This brings the maximum demand up to about 137 A. The nearest HRC fuse size for this load is 160 A so this is the fuse chosen for the existing distribution board.

The feeder cable to the kitchen distribution board can now be sized. The cable is clipped direct and could be subject to overload, so it has to be sized to the fuse rating I_n. Looking at Table 3.3 column 3, a 70 mm^2 PVCSWA & PVC armoured cable can be used having an I_{tab} of 192 A.

The next step would be to choose the fuse ratings for the loads on the kitchen distribution board and make sure these will discriminate

Table 2.2   *Tabulated figures for the worked example showing current ratings*

Circuit item	Load kW	Diversity % f.l.c.	Revised Load	Red phase amperes	Yellow phase amperes	Blue phase amperes
Three-phase loads						
1 Dishwasher	25	100%	25	34.78	34.78	34.78
2 Oven	12	80%	9.6	13.36	13.36	13.36
3 Fryer	11	60%	6.6	9.18	9.18	9.18
4 Hot plate	9	60%	5.4	7.51	7.51	7.51
5 Hot plate	9	60%	5.4	7.51	7.51	7.51
6 I. D. fan	3	100%	3.0	6.5	6.5	6.5
7 F. D. fan	2.2	100%	2.2	5.0	5.0	5.0
Total three-phase load				83.84	83.84	83.84
Single-phase loads						
8 Waste disp	1.5	60%	0.9	3.75		
9 Microwave	3	60%	1.8		7.5	
10 Freezer	1	60%	0.6			2.5
11 Freezer	1	60%	0.6		2.5	
12 Lighting	6 A	90%	5.4 A	5.4		
13 Boiler panel	6 A	100%	6 A			6.0
14 Ring circuit	32 A	32 A	32 A			32.0
Total load				92.99	93.84	124.34

with the 160 A fuse installed in the factory distribution board. Next determine which circuits are protected only against short-circuit current i.e. resistive fixed loads and those which are protected against overload. Diversity is not applied to the final circuit, since when it is in use it will take the current required by the load, so the diversified loads in Table 2.2 cannot be used for sizing the fuses.

The fuse ratings are listed in Table 2.3. Only circuits 6, 7, 8 and 14 are protected against overload, all the others are fixed loads. Although the freezers have a compressor in them they work in a closed circuit where the load is fixed, they also have internal cutouts should the refrigeration cycle go faulty.

If all the circuits were grouped together the derating factor would make the circuit conductors too large for the terminals in the equipment. So the circuits were divided to improve the derating factor. Circuits 1, 2 and 12 were fed individually, the remainder were grouped as follows: circuits 3, 4, 5 and 9; circuits 10, 11 and 14; circuits 6, 7, 8 and 13. In practice the groupings would be arranged to suit the location of the equipment.

The installation will not be in contact with thermal insulation, nor will it be installed in areas with an ambient temperature in excess of 30°C, so if we use HRC fuses for protection the only derating factor is for grouping.

For circuits 1, 2 and 12 the cable size is determined from the design current and is shown in Table 2.3.

Circuits 3, 4, 5 and 9 are fixed loads and are, therefore, only being protected against short-circuit current. There are four circuits grouped so the derating factor from Table 3.4 is 0.65. Since simultaneous overload may be assumed not to occur the current-carrying capacity required for the cable can be calculated as follows:

First, calculation for fryer.

$$I_{t_1} = \frac{I_b}{G \times A \times T \times S} \quad \text{so that} \quad I_{t_1} = \frac{15.3}{0.65 \times 1 \times 1 \times 1} = 23.54 \text{ A}$$

Second calculation:

$$I_{t_2} \geq \frac{1}{A \times T} \sqrt{I_n^2 - 0.48 I_b^2 \left(\frac{1 - G^2}{G^2} \right)} \text{ A}$$

$$I_{t_2} \geq \frac{1}{1 \times 1} \sqrt{20^2 - 0.48 \times 15.3 \left(\frac{1 - 0.65^2}{0.65^2} \right)} = 19.75 \text{A}$$

The cable has to be sized to the larger of I_{t_1} or I_{t_2}, which from above is I_{t_1}.

The cable size required from IEE Table 4D1A is 4 mm^2 with an I_{tab} of 28 A in conduit. The other cable sizes for those in the group are given in Table 2.3.

With regards to circuits 6, 7, 8 and 13, circuits 6 and 7 are both for motors so the cables should be sized to the maximum overload setting in the starter, which is 9 A. The fuse is only providing fault current protection.

Simultaneous overload can occur so the cables have to be sized to the overload protective device. Again the derating factor will be 0.65.

$$I_t = \frac{I_n}{G} = \frac{9\text{A}}{0.65} = 13.85\text{A}$$

A 1.5 mm^2 cable will be suitable and is listed in Table 2.3.

This leaves circuits 10, 11 and 14 grouped together. From inspection of the fuse sizes it will be obvious that 1.5 mm^2 will be suitable

Table 2.3 Tabulated figures for the worked example showing protection, cable sizes and voltage drops

Equipment	Circuit	Load kW	Phase	Type	Design current Amps	Fuse rating Amps	G, A, T, S factors	I_{tab} Amps	Cable size sq.mm	I_{tab} Amps	mV/A/m	length m	Volt drop Volts
Dishwater	1	25	3 phase	Fixed load	34.78	35 A	–	34.78	6	36	3.4	18	4.01
Oven	2	12	3 phase	Fixed load	16.69	20 A	–	16.69	2.5	21	15	18	4.51
Fryer	3	11	3 phase	Fixed load	15.3	20 A	0.65	23.54	4	28	9.5	18	2.62
Hotplate	4	9	3 phase	Fixed load	12.52	16 A	0.65	19.26	2.5	21	15	18	3.38
Hotplate	5	9	3 phase	Fixed load	12.52	16 A	0.65	19.26	2.5	21	15	18	3.38
I. D. Fan	6	3	3 phase	Motor	6.5	16 A	0.65	13.85	1.5	15.5	25	18	2.93
F. D. Fan	7	2.2	3 phase	Motor	5	16 A	0.65	13.85	1.5	15.5	25	18	2.25
Waste disposal	8	1.5	1 phase	Variable	6.25	10 A	0.65	15.38	1.5	17.5	29	18	3.26
Microwave	9	3	1 phase	Fixed load	12.5	16 A	0.65	19.23	2.5	24	18	18	4.05
Freezer	10	1	1 phase	Fixed load	4.2	6 A	0.7	6	1.5	17.5	29	18	2.19
Freezer	11	1	1 phase	Fixed load	4.2	6 A	0.7	6	1.5	17.5	29	18	2.19
Lighting	12	6 A	1 phase	Fixed load	6	6 A	–	6	1.5	17.5	29	18	3.13
Boiler control	13	6 A	1 phase	Control	6	6 A	0.65	9.23	1.5	17.5	29	18	3.13
Ring circuit	14	32 A	1 phase	Socket	32	32 A	0.7	30.63	4	32	11	38	3.34

for circuits 10 and 11, but the current-carrying capacity of the cables for circuit 14, which is a ring circuit will have to be calculated. The derating factor for three circuits is 0.7.

$$I_t = \frac{I_n \times 0.67}{G} = \frac{32\text{A} \times 0.67}{0.7} = 30.63\text{A}$$

From IEE Table 401 A the cable size required for single-core cable in conduit is $4\,\text{mm}^2$.

Voltage drop Having sized the cables for the load and derating factors the next step is to check that the cables are suitable for voltage drop since this can affect the cable sizes chosen.

The equipment supplied from final circuits will only receive a certain size of cable, so it pays to determine the final circuit cable sizes first, then work back from the final circuits to the supply point for determining the voltage drop.

The voltage drop is worked out by using the formula given earlier in this chapter and it is shown for each circuit in Table 2.3.

As explained earlier, the voltage drop in a ring circuit, or any other socket-outlet circuit for that matter, can only be an approximation.

In the case of the ring circuit in the kitchen, the calculation is as follows.

Half length of ring circuit is 19 metres; half load is 16 A; mV/A/m for cable size required is 11.

$$V_d = \frac{19 \times 16 \times 11}{1000} = 3.34\,\text{V}$$

It is now necessary to determine the voltage drop in the cable from the factory distribution board. Length of cable 60 m, maximum demand from Table 2.2 is 124.34 A plus 10% future allowance gives 137 A. The mV/A/m for $70\,\text{mm}^2$ cable is $0.57\,\Omega/\text{A}/\text{m}$ three-phase.

$$\text{Three-phase voltage drop} = \frac{60\text{m} \times 137\text{A} \times 0.57\Omega}{1000} = 4.69\,\text{V}$$

The single phase voltage will be $4.69 \div \sqrt{3} = 2.71\,\text{V}$

The voltage drop up to the factory distribution board is 6 V three-phase and 3.46 V single phase (6 ÷ 1.732).

The voltage drop up to the factory distribution board must now be added to the voltage drop in the feeder cable to the kitchen and the total added to the voltage drop in each final circuit, to ensure that the single phase voltage and three phase voltage drops do not exceed 4% i.e. 9.6 V single-phase or 16.6 V three-phase.

The maximum three-phase voltage drop allowed in a final circuit will be:

$$16.6\,V - (6 + 4.69) = 5.91\,V$$

The maximum single-phase voltage drop allowed in a final circuit will be:

$$9.6\,V - (3.46 + 2.71) = 3.43\,V$$

The only circuit that does not comply is number 9 and the cable size will have to be increased to 4 mm² with a mV/A/m of 11. The revised voltage drop will be:

$$V_{d\,circ\,9} = \frac{18\ m \times 12.5 \times 11}{1000} = 2.475\ V$$

This is now less than the maximum allowed of 3.43 V.

Short-circuit current   The maximum and minimum short-circuit current have to be established, the first to enable equipment with the correct withstand capacity to be selected and the latter to enable cables to be checked.

The necessary information to complete the calculations is not available in the Wiring Regulations but can be obtained from cable and fuse manufacturers. For convenience the following information has been extracted from the Handbook on the IEE Wiring Regulations by Trevor E. Marks: the impedance of the 70 mm² PVCSWA & PVC cable at 20°C is 0.279 Ω per 1000 m; the impedance at 115°C is 0.378 Ω per 1000 m; the I^2t let through for a 35 A fuse is 5300 A² sec and for a 16 A fuse 550 A² sec.

In a three-phase system the maximum value is determined by taking the impedance of one phase only up to the point being considered and then dividing it into the phase to neutral voltage.

The maximum fault current at the factory distribution board I_p is 5518 A. The minimum impedance of one phase will therefore be:

$$Z_P = \frac{240}{5518}\ \Omega = 0.0435\ \Omega$$

To find the maximum fault level at the kitchen distribution board the impedance of one phase of the feeder cable is added to the Z_p just obtained. The impedance is taken at 20°C. The impedance of 70 mm^2 PVCSWA & PVC cable at 20°C is

$$0.000279 \times 60\,m = 0.01674\,\Omega$$

Total impedance at kitchen distribution board is 0.0435 + 0.01674 = 0.06 Ω.

The fault level will be 240 V divided by 0.06 ohm = 4000 A.

This is well within the capability of HRC fuses, but if miniature circuit breakers were being used, it would have to be checked that they had a breaking capacity in excess of 4 kA.

The next step is to check that the cables are protected. The minimum impedance up to the factory distribution board will be the phase voltage divided by the single-phase fault current.

$$Z_{pn} = \frac{230}{3380}\,\Omega = 0.068\,\Omega$$

The impedance at 115°C for the 70 mm^2 phase and neutral conductors is 0.000378 Ω/m.

The phase-neutral impedance up to kitchen distribution board is

$$0.068 + (60\,m \times 0.000378 \times 2) = 0.113\,\Omega$$

The minimum fault current will be:

$$I_{pn} = \frac{230}{0.113}\,A = 2035A$$

The breaking capacity of a BS 88 Part 2 HBC fuse is 80 kA which exceeds the calculated minimum short-circuit current, therefore, any circuit in which the current-carrying capacity of the conductors exceeds the rating of the protective device, can be assumed to be protected against short-circuit current (IEE Regulation 434-03-02).

Looking at Table 2.3 only circuits 6 and 7 do not have a current-carrying capacity equal to or greater than the rating of the protective device and do not, therefore, comply with IEE Regulation 434-03-02. However, for the purposes of illustration in this example it will be assumed that circuit 1 and 6 do not comply with IEE Regulation 434-03-02.

Circuit number 1
This is a TP&N circuit so the minimum short-circuit current will be
between phase and neutral. The fuse size is 35 A, Cable size 6 mm²,
average temperature 115°C under fault conditions.
Z_{pn} up to the kitchen is 0.1164 Ω.

Impedance of 6 mm² phase and neutral

$$= \frac{18 \text{ m} \times 2 \times 3.08 \times 1.38}{1000} = 0.153 \Omega$$

Total impedance Z_{pn} = 0.153 + 0.1164 = 0.269 Ω

Fault current $I_{pn} = \frac{230}{0.269} \text{ A} = 855 \text{ A}$

Maximum time allowed by IEE Regulation 434-03-03

$$t = \frac{k^2 S^2}{I^2} \text{ secs} = \frac{115^2 \times 6^2}{855^2} \text{ secs} = 0.65 \text{ secs}$$

It is now necessary to look at the fuse characteristic for the 35 A fuse
to determine the actual disconnection time. The disconnection time
is less than 0.1 secs and therefore the $I^2 t$ energy let through from the
manufacturer has to be compared with $k^2 S^2$ in accordance with the
note to IEE Regulation 434-03-03.
   Total $I^2 t$ for 35 A fuse is 5300, which is exceeded by k^2 alone, i.e.
13 225, so the cable is protected.

Circuit number 6
This is a three-wire three-phase circuit so the minimum short-circuit
current will be with a phase-to-phase fault.
   The minimum Z_p up to the kitchen distribution board:

$$Z_p = 0.0435 + (60 \text{ m} \times 0.000378) = 0.066 \Omega$$

Circuit impedance for 1.5 mm² cable = 18 m × 0.0121 × 1.38
$$= 0.3 \Omega$$

Total impedance of one phase = 0.3 + 0.066 = 0.37 Ω

Short circuit current $I_{pp} = \frac{V_L}{2 \times Z_p} = \frac{415}{2 \times 0.37} = 561 \text{ A}$

Maximum time allowed by regulations for disconnection

$$t = \frac{115^2 \times 1.5^2}{561^2} \text{ secs} = 0.095 \text{ secs}$$

Actual disconnection time of 16 A fuse is less than 0.1 secs. The I^2t energy let through for a 16 A fuse is 550 A^{2}s. This will be less than k^2S^2 so the cable is protected.

Protective conductors   To determine that the protective conductors are thermally protected and of the correct size it is necessary to know the phase earth loop impedance.

The following additional information has been extracted from the Handbook on the IEE Wiring Regulations by Trevor E. Marks: Design values for: phase earth loop impedance of 4 × 70 mm^2 PVCSWA & PVC cable 1.89 Ω/1000 m, phase conductor impedance 0.378 Ω/1000 m at 115°C, gross area of armour 138 mm^2, impedance of 20 mm heavy gauge steel conduit 0.005 875 Ω/m, cross-sectional area 107 mm^2 and I^2t for 35 A HRC fuse 5300 A^{2}secs.

Feeder cable to kitchen distribution board

$$Z_{\text{inst}} = \frac{60 \text{ m} \times 1.89}{1000} = 0.1134 \, \Omega$$

$$Z_S = Z_E + Z_{\text{inst}} = 0.19 + 0.1134 = 0.303 \, \Omega$$

$$\text{Fault current } I_f = \frac{U_0}{Z_S} \text{ A} = \frac{230}{0.303} \text{ A} = 759 \text{A}$$

From the IEE Wiring Regulations the disconnection time from Appendix 3 BS 88 Part 2 fuse characteristic is 8.5 secs for 160 A fuse and the value of k from Table 54D is 51.

Minimum cross sectional area of protective conductor is given by

$$S = \frac{\sqrt{I_f^2 t}}{k} = \frac{\sqrt{759^2 \times 8.5}}{51} = 43.39 \text{ mm}^2$$

The area of the armour is 138 mm^2 and exceeds this value, so the armour is thermally protected. However, as will be shown later the disconnection time will be unacceptable.

Final circuits

Circuit number 1

Before the final circuit can be calculated the size of conduit has to be worked out. Circuit number 1 will be worked out as an example.

It is planned that there will be a maximum of two bends in every 3 metres of the conduit, with draw-in boxes every 3 metres.

Size of cable $6\,mm^2$, number of conductors 4.

From Table 3.5B, factor for $6\,mm^2$ cable is 58. Total factors $58 \times 4 = 232$.

Now look at Table 3.5A under the columns for two bends and opposite the length of run for 3 metres. The factor under the 20 mm column is 233, so a 20 mm conduit can be used.

$$Z_{inst} = 18((0.00308 \times 1.38) + 0.005\,875) = 0.182\,\Omega$$

Z_E is now required and this will be the previously calculated Z_S at the bus bars of the kitchen distribution board.

$$Z_s = Z_E + Z_{inst} = 0.303 + 0.182 = 0.485\,\Omega$$

$$I_f = \frac{230}{0.485} = 474\,A$$

The disconnection time is less than 0.1 secs therefore $I^2t \leqslant k^2S^2$.

I^2t for the 35 A fuse is $5300\,A^2$secs
$k^2S^2 = 47^2 \times 107^2 = 25\,290\,841\,A^2$secs

Thermal capacity of conduit is more than satisfactory. It will generally be found that the thermal capacity of steel conduit will always be satisfactory and due to its large cross-sectional area it is an excellent protective conductor providing it is used in the correct locations and good workmanship is used. It is a requirement of Section 3 of the IEE Wiring Regulations that any protective measure used shall remain effective during the life of the installation. Installing conduits in areas where there will be a gradual deterioration of the conduit during the life of the installation is not complying with the IEE Wiring Regulations.

Protection against indirect contact   The earth loop impedance has already been worked out for the protective conductors and only need checking against the tables in the IEE Wiring Regulations.

Feeder cable to kitchen $Z_S = 0.303\,\Omega$.

Maximum Z_s allowed from IEE Table 41D is 0.27 Ω. The calculated earth loop impedance therefore is not suitable to protect indirect contact.

There are three options open: the first to increase the size of feeder cable so that the earth loop impedance falls below 0.27 Ω, the second is to install a separate protective conductor to reduce the earth loop impedance, the third option is to install a protective conductor which will reduce the earth loop impedance and at the same time comply with IEE Table 41C; this would have the advantage of not having to install local main equipotential bonding conductors from the distribution board in the kitchen due to a socket outlet circuit being fed from the same distribution board as stationary equipment. Unfortunately this last option, i.e. using IEE Table 41C, is not available since the impedance from the main earth bar to the factory distribution board is unknown.

For the purposes of this example it will be assumed that the main earth bar is at the factory distribution board.

Since it is now proposed to install a protective conductor in parallel with the armouring of the cable it is simpler to ignore the armouring of the cable although it will still be in the phase-earth loop.

To use IEE Table 41C it is necessary to know the earth loop impedance for the protective device for 5 secs disconnection time. From IEE Table 41D this is 0.27 Ω.

The maximum impedance allowed for the protective conductor from IEE Regulation 413-02-13 is:

$$\frac{50 \times Z_S}{U_0} = \frac{50 \times 0.27}{230} = 0.0587 \; \Omega$$

The impedance per metre should not exceed $0.0587 \div 60 = 0.000\,938\,\Omega$ If this protective conductor is installed two cable diameters away from the feeder cable it can be assumed that the temperature of the protective conductor is 30°C. The resistance correction factor for design when the conductor is operating at 30°C is 1.30.

Resistance of a 25 mm^2 copper cable

$$\frac{0.727 \times 1.30}{1000} = 0.000\,945\,1\Omega \text{ per metre}$$

This is higher than that required so a 35 mm^2 conductor will be needed.

Total resistance would then be $0.000\,524 \times 1.30 \times 60\,\text{m} = 0.0409\,\Omega$. It must now be checked that the revised earth loop impedance will cause disconnection within 5 secs.

The impedance of the phase conductor from the factory to the kitchen is

$$Z_p = 60\,\text{m} \times 0.000\,378\,\Omega = 0.0227\,\Omega \text{ at } 115°\text{C}$$

The impedance of the $35\,\text{mm}^2$ copper protective conductor is Z_{cpc} is $0.0409\,\Omega$

$$Z_s = Z_E + Z_{inst}$$
$$Z_s = 0.19 + (0.0227 + 0.0409) = 0.25\,\Omega$$

This now is less than the Wiring Regulations Requirement of $0.27\,\Omega$ The earth loop impedance will be less than $0.25\,\Omega$ since the cable armour will be contributing to lowering the value of Z_S up to the kitchen distribution board.

The thermal capacity of the $35\,\text{mm}^2$ protective conductor should now be checked. On inspection its area is half that of the phase conductor, so it complies with IEE Table 54G and is therefore satisfactory.

Circuit number 1
The calculated earth loop impedance for circuit number 1 from above is $0.485\,\Omega$, the fuse size protecting the circuit is 35 A. The maximum earth loop impedance is not given in the IEE Wiring Regulations but Z_s allowed for a 40 A fuse is $1.41\,\Omega$. This is greater than $0.485\,\Omega$ so the circuit is protected against indirect contact even without the installation of the additional protective conductor to the kitchen distribution board.

General   Only a few circuits have been analysed and calculated in detail due to lack of space, however, the calculations for each circuit would be carried out in a similar way to those illustrated.

Design by computer

The above example will have illustrated how complicated and long winded calculations can become when numerous circuits have to be worked out.

Fortunately with the rapid advance and spread of computing equipment, software designers have seized the opportunity of producing programmes which reduce the time taken to design an installation. Care is needed in choosing such a programme that it takes into account all the parameters. In general it can only do this if it puts questions to the operator to make sure that the correct formula will be used. Unfortunately many programmes don't do this and just give the answer instead of showing the full calculations carried out, thus depriving the engineer of the opportunity of checking the result.

The designer can expect to find a number of benefits from using computers for installation design.

(1) The speed at which calculation is carried out will clearly be much faster than with manual methods. Even with use of a calculator in the hands of an experienced designer, manual methods can take up to 20 minutes per circuit for all aspects to be properly assessed. The computer will be able to do the work in a fraction of the time.

(2) Nearly all the work of looking at tables can be eliminated. The key facts about cable ratings, volt drops, the fuse and earth fault loop impedance data, and even the manufacturers' characteristics on devices and equipment, can all be stored in the memory of the computer. Many electrical manufacturers are able to supply product data in computer format for the popular software packages, the disk contents being fed straight into the computer.

(3) Repetition of calculations can easily be carried out. In settling the question of the best location of a distribution board, or whether it would be more economic to avoid grouping by fixing cables separately, it is possible to quickly recalculate to decide the optimum condition. With manual methods, the work would be considerable and it is probable that only one set of calculations would be made.

(4) Potentially the computer method will be more accurate. This is of course fundamentally dependent upon the quality of the software, but provided this is satisfactory, the risk of human error, present in the manual methods, is avoided. Again speed of calculation is of advantage, as the simplified formulae sometimes used in manual methods can be avoided. The computer will be able to use the most appropriate and accurate method of calculation for every single circuit.

(5) Design data for the installation can be produced by computer print-out with little extra effort. The manually produced design data sheet which may take some time to prepare without computers, is one of the tasks that can be done more quickly.

Any designer considering using computers for the first time should carefully assess the various software programs which are available, in particular that the program is handling the formulae correctly and is asking the correct type of questions of the user before proceeding with the calculation; this does mean that you must fully understand how to apply the regulations. The advice of other users of particular packages can be a source of information and potential users need to be sure the software on offer is appropriate for their purposes. Because computers and software are developing so rapidly, it is as well to review a number of different software products. Care should also be taken over the selection of the computers themselves, the hardware. Use of existing office machines may not be appropriate and advice from the software supplier should be heeded so as to get the best performance from the program.

Although the use of computers in the installation design process has many advantages, some of which are referred to above, users should be aware of some issues relevant to the constraints which can occur.

(1) Electrical installation design still requires full understanding of the fundamentals of electrical engineering principles. Because the computer is very quick and accurate at calculation, it can sometimes mistakenly be thought that it will make the decisions as well. This is not so.

(2) As with all computers, proper systematic arrangements must be made to save data on a regular basis, and to back up all important work. If the work of the design office is likely to affect production, the provision of more than one computer may be appropriate to avoid delays in the event of machine failure. Once designs are complete, secure storage of duplicate sets of design data is a must.

(3) Any software should be purchased from reputable companies, and there are two main reasons for this. Firstly, future technical support and advice relating to the design package may well be needed and it is useful to deal with a company which is likely to remain in business. Secondly, as electrical wiring and other regulations develop, the software could well require updating to reflect the changes, and reputable firms are likely to provide for these developments in their range of software products or services provided.

Figures 2.18 to 2.21 are taken from the outputs of a typical software package supplied for electrical installation design.

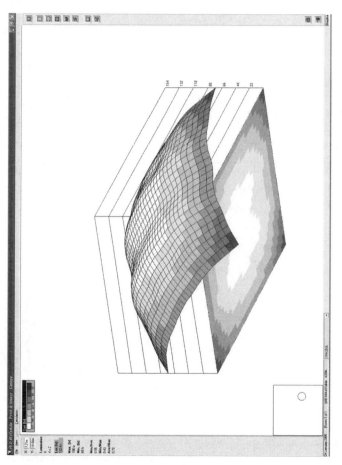

Fig. 2.18 Computer results from lumen calculation. The results are available in tabular and diagrammatic form on either the screen or printer. This example is for a room of 19 × 7.2 m two rows of four luminaires. The print shows the point to point light levels (W.T. Parker Ltd)

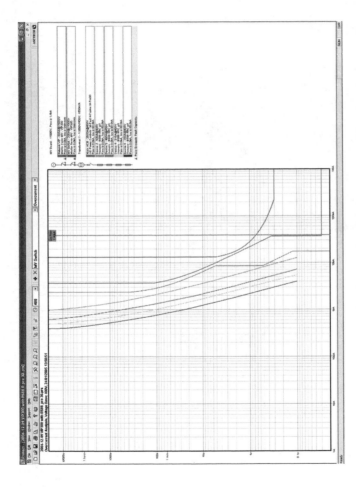

Fig. 2.19 *The power of computing is well illustrated in this graphical representation of the characteristics of protective devices in an installation (see Fig. 2.10). By constructing the protection curves it is possible to ensure that there is discrimination and it is a simple matter to revise the specification, if needed, at the design stage to ensure that discrimination between devices is maintained (W.T. Parker Ltd)*

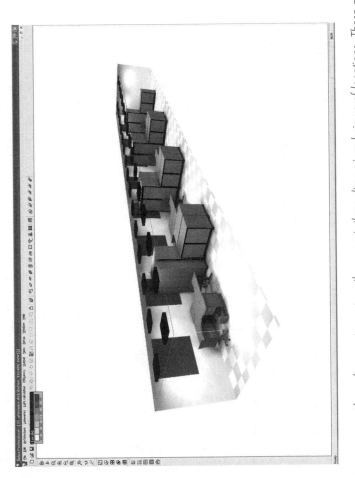

Fig. 2.20 *Computing power with modern equipment easily supports three-dimensional views of locations. These are being used in this office installation to plan and specify the lighting scheme. This illustration is a screen print of the room layout. The two subsequent views on pp. 88 and 89 show the luminaire locations and calculation and finally there is a graphical depiction of light levels. The scheme used 30 Zumtobel fittings and delivered 105,000 lumens (W.T. Parker Ltd)*

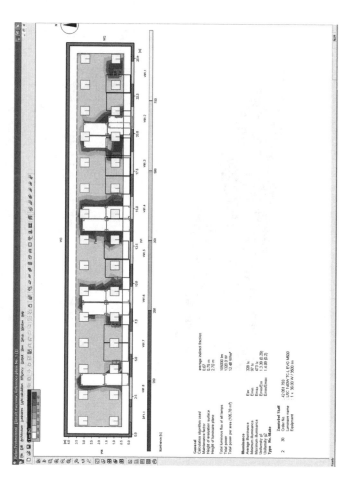

Fig. 2.20 *Continued*

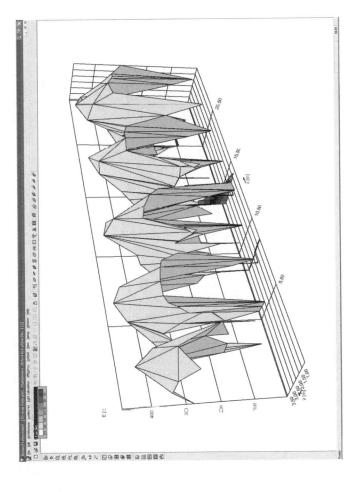

Fig. 2.20 *Continued*

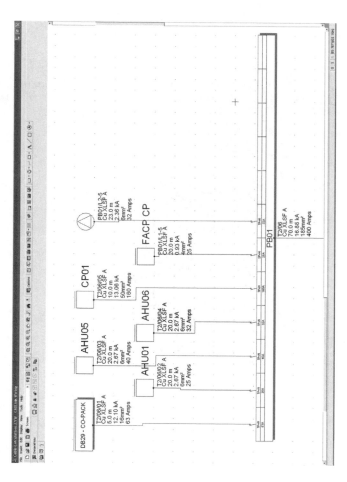

Fig. 2.21   *With computer-aided circuit design, full details of distribution arrangements are entered by the designer, and held in the computer memory. A typical 3-phase installation arrangement is shown with distribution boards and cable details shown. Systematic recording of design assumptions is an essential feature of the design process and the information can be presented in a number of different ways. The following tables show computer prints. The view on p. 91 is a table of distribution board data and the example on p. 92 shows full details for a particular cable (W.T. Parker Ltd)*

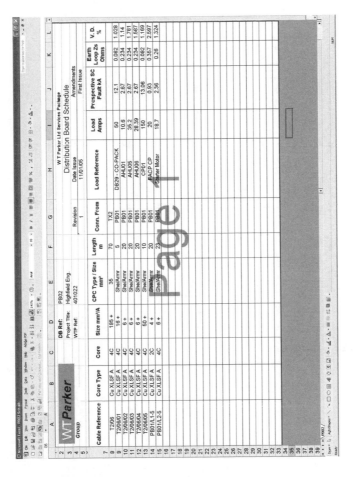

The distribution board schedule shows a Microsoft Excel spreadsheet titled "WT Parker Ltd Services Package — Distribution Board Schedule".

Cable Reference	Core Type	Core	Size mm²/A	CPC Type / Size mm²	Length m	Conn. From	Load Reference	Load Amps	Prospective SC Fault kA	Earth Loop Zs Ohms	V. D. %
T206	Cu XLSF A	4C	185 +	35	70	TX2					
T206/01	Cu XLSF A	4C	16 +	She/Armr	5	PB01	DB29 - CO-PACK	50	12.1	0.082	1.028
T206/02	Cu XLSF A	4C	6 +	She/Armr	20	PB01	AHU01	10.6	2.67	0.234	1.14
T206/03	Cu XLSF A	4C	6 +	She/Armr	20	PB01	AHU05	35.2	2.67	0.234	1.781
T206/04	Cu XLSF A	4C	6 +	She/Armr	20	PB01	AHU06	28.39	2.67	0.234	1.587
T206/05	Cu XLSF A	4C	50 +	She/Armr	10	PB01	CP01	150	13.06	0.082	1.169
PB01/1.1-5	Cu XLSF A	2C	4 +	She/Armr	20	PB01	FACP CP	20	0.93	0.357	2.597
PB01/1.2-5	Cu XLSF A	4C	6 +	She/Armr	23	PB01	Starter Motor	18.7	2.36	0.26	1.324

DB Ref: PB02
Project Title: Highfield Eng.
WTP Ref: 401022

Revision 1

Date Issue 11/01/05

Amendments
First Issue

Fig. 2.21 Continued

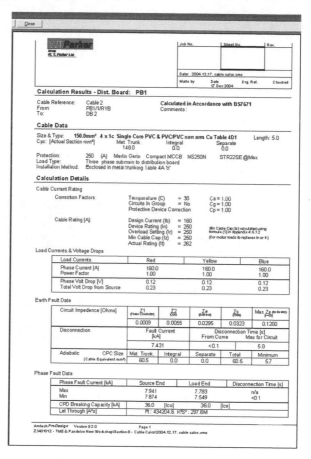

Fig. 2.21   *Continued*

3 Tables from IEE Regulations for Electrical Installations

This chapter includes some of the tables from the IEE Regulations for Electrical Installations. The tables include those giving current-carrying capacities and voltage drops for cables and flexible cords, as well as grouping and other derating factors. The chapter also includes tables of cable capacities for conduit and trunking, and cable resistance.

As only a few tables are included here, it is necessary to consult the IEE Regulations for Electrical Installations to get full details for all types of cables. Information on the method of using the IEE Tables to carry out the various design calculations necessary is given in Chapter 2 of this book.

The tables reproduced are listed below.

Table number	Page	IEE table number	Title
3.1	94	53A	Sizes of fuse elements for BS 3036 Fuses
3.2	95–6	4D2A, B	Current-carrying capacities and voltage drops for PVC-insulated cables (non-armoured)
3.3	97–8	4D4A, B	Current-carrying capacities and voltage drops for PVC-insulated cables (armoured)
3.4	99	4B1	Correction factors for grouping
3.5A	100		Conduit factors
3.5B	101		Cable factors for conduit
3.6	101		Cable factors for trunking
3.7	102		Cable resistance

Table 3.1   *Sizes of fuse elements of plain or tinned copper wire for use in semi-enclosed fuses (IEE Table 53A) (To BS 3036)*

Nominal current of fuse (A)	Nominal diameter of wire (mm)
3.0	0.15
5.0	0.20
10.0	0.35
15.0	0.50
20.0	0.60
25.0	0.75
30.0	0.85
45.0	1.25
60.0	1.53
80.0	1.80
100.0	2.00

Note: The manufacturers should be consulted as to the correct size of fusewire to use in their fuse holders. Using a larger cross-sectional area of fusewire than recommended by the manufacturer will lead to burning of the fuse holder.

Table 3.2 Current-carrying capacities and voltage drops for multicore PVC-insulated cables, non-armoured (copper conductors) (IEE Table 4D2A) (BS 6004, BS 6346, BS 7629)

Ambient temperature: 30°C
Conductor operating temperature: 70°C

Current-carrying capacities (amperes):

Conductor cross-sectional area	Reference Method 4 (enclosed in an insulated wall, etc.)		Reference Method 3 (enclosed in conduit on a wall or ceiling, or in trunking)		Reference Method 1 (clipped direct)		Reference Method 11 (on a perforated cable tray), or Reference Method 13 (free air)	
	1 two-core cable* single-phase a.c. or d.c.	1 three-core cable*, or 1 four-core cable, three phase a.c.	1 two-core cable*, single phase a.c or d.c	1 three-core cable*, or 1 four-core cable, three phase a.c.	1 two-core cable* single-phase a.c. or d.c.	1 three-core cable*, or 1 four-core cable, three-phase a.c.	1 two-core cable*, single-phase a.c. or d.c.	1 three-core cable*, or 1 four-core cable three-phase a.c.
1	2	3	4	5	6	7	8	9
mm²	A	A	A	A	A	A	A	A
1	11	10	13	11.5	15	13.5	17	14.5
1.5	14	13	16.5	15	19.5	17.5	22	18.5
2.5	18.5	17.5	23	20	27	24	30	25
4	25	23	30	27	36	32	40	34
6	32	29	38	34	46	41	51	43
10	43	39	52	46	63	57	70	60
16	57	52	69	62	85	76	94	80
25	75	68	90	80	112	96	119	101
35	92	83	111	99	138	119	148	126
50	110	99	133	118	168	144	180	153
70	136	125	168	149	213	184	232	196
95	167	150	201	179	258	223	282	238
120	192	172	232	206	299	259	328	276
150	219	196	258	225	344	299	379	319
185	248	223	294	255	392	341	434	364
240	291	261	344	297	461	403	514	430
300	334	298	394	339	530	464	593	497
400	–	–	470	402	634	557	715	597

Table 3.2 (continued) (IEE Table 4D2B)

Volt drops (mV/A/m)

Conductor cross-sectional area (1)	Two-core cable d.c. (2)	Two-core cable single phase a.c. (3)			Three- or four-core cable three phase a.c. (4)		
mm²	mV	mV			mV		
1	44	44			38		
1.5	29	29			25		
2.5	18	18			15		
4	11	11			9.5		
6	7.3	7.3			6.4		
10	4.4	4.4			3.8		
16	2.8	2.8			2.4		
	r	r	x	z	r	x	z
25	1.75	1.75	0.170	1.75	1.50	0.145	1.50
35	1.25	1.25	0.165	1.25	1.10	0.145	1.10
50	0.93	0.93	0.165	0.94	0.80	0.140	0.81
70	0.63	0.63	0.160	0.65	0.55	0.140	0.57
95	0.46	0.47	0.155	0.50	0.41	0.135	0.43
120	0.36	0.38	0.155	0.41	0.33	0.135	0.35
150	0.29	0.30	0.155	0.34	0.26	0.130	0.29
185	0.23	0.25	0.150	0.29	0.21	0.130	0.25
240	0.180	0.190	0.150	0.24	0.165	0.130	0.21
300	0.145	0.155	0.145	0.21	0.135	0.130	0.185
400	0.105	0.115	0.145	0.185	0.100	0.125	0.160

* With or without protective conductor.

Circular conductors are assumed for sizes up to and including 16 mm². Values for larger sizes relate to shaped conductors and may safely be applied to circular conductors.

Note: Where the conductor is to be protected by a semi-enclosed fuse to BS 3036, see item 6.2 of IEE Appendix 4. Cables to BS 7629 are rated for a conductor operating temperature of 70°C and are therefore included in this table, although the material used for the cable insulation is not PVC.

Table 3.3 Current-carrying capacities and voltage drops for multicore armoured PVC-insulated cables (copper conductors) (IEE Table 4D4A)

Current-carrying capacities (amperes):

Ambient temperature: 30°C
Conductor operating temperature: 70°C

Conductor cross-sectional	Reference Method 1 (clipped direct)		Reference Method 11 (on a perforated horizontal cable tray, or Reference Method 13 (free air))	
	1 two-core cable, single-phase a.c. or d.c.	1 three- or four-core cable, three-phase a.c. or d.c.	1 two-core cable, single-phase a.c. or d.c.	1 three- or four-core cable, three-phase a.c.
1	2	3	4	5
mm²	A	A	A	A
1.5	21	18	22	19
2.5	28	25	31	26
4	38	33	41	35
6	49	42	53	45
10	67	58	72	62
16	89	77	97	83
25	118	102	128	110
35	145	125	157	135
50	175	151	190	163
70	222	192	241	207
95	269	231	291	251
120	310	267	336	290
150	356	306	386	332
185	405	348	439	378
240	476	409	516	445
300	547	469	592	510
400	621	540	683	590

Table 3.3 (continued) (IEE Table 4D4B)

Volt drops (mV/A/m)

Conductor cross-sectional area 1	Two-core cable d.c. 2	Two-core cable single phase a.c. 3			Three- or four-core cable three phase a.c. 4		
mm²	mV	mV			mV		
1.5	29	29			25		
2.5	18	18			15		
4	11	11			9.5		
6	7.3	7.3			6.4		
10	4.4	4.4			3.8		
16	2.8	2.8			2.4		
		r	x	z	r	x	z
25	1.75	1.75	0.170	1.75	1.50	0.145	1.50
35	1.25	1.25	0.165	1.25	1.10	0.145	1.10
50	0.93	0.93	0.165	0.94	0.80	0.140	0.81
70	0.63	0.63	0.160	0.65	0.55	0.140	0.57
95	0.46	0.47	0.155	0.50	0.41	0.135	0.43
120	0.36	0.38	0.155	0.41	0.33	0.135	0.35
150	0.29	0.30	0.155	0.34	0.26	0.130	0.29
185	0.23	0.25	0.150	0.29	0.21	0.130	0.25
240	0.180	0.190	0.150	0.24	0.165	0.130	0.21
300	0.145	0.155	0.145	0.21	0.135	0.130	0.185
400	0.105	0.115	0.145	0.185	0.100	0.125	0.160

Note: Where the conductor is to be protected by a semi-enclosed fuse to BS 3036, see item 6.2 of IEE Appendix 4.

Table 3.4 Correction factors for groups of more than one circuit of single-core cables, or more than one multicore cable (Extract from IEE Table 4B1)

Reference method of installation (see Table 9A)		Correction factor (C_g)													
		Number of circuits or multicore cables													
		2	3	4	5	6	7	8	9	10	12	14	16	18	20
Enclosed (Method 3 or 4) or bunched and clipped direct to a non-metallic surface (Method 1)		0.80	0.70	0.65	0.60	0.57	0.54	0.52	0.50	0.48	0.45	0.43	0.41	0.39	0.38
Single layer clipped to a non-metallic surface (Method 1)	Touching	0.85	0.79	0.75	0.73	0.72	0.72	0.71	0.70	–	–	–	–	–	–
	Spaced*	0.94	0.90	0.90	0.90	0.90	0.90	0.90	0.90	0.90	0.90	0.90	0.90	0.90	0.90
Single layer multicore on a perforated metal cable tray, vertical or horizontal (Method 11)	Touching	0.86	0.81	0.77	0.75	0.74	0.73	0.73	0.72	0.71	0.70	–	–	–	–
	Spaced*†	0.91	0.89	0.88	0.87	0.87	–	–	–	–	–	–	–	–	–
Single layer single-core on a perforated metal cable tray, touching (Method 11)	Horizontal	0.90	0.85	–	–	–	–	–	–	–	–	–	–	–	–
	Vertical	0.85	–	–	–	–	–	–	–	–	–	–	–	–	–
Single layer multicore touching on ladder supports		0.86	0.82	0.80	0.79	0.78	0.78	0.78	0.77	–	–	–	–	–	–

* 'Spaced' means a clearance between adjacent surfaces of at least one cable diameter (D_e). Where the horizontal clearances between adjacent cables exceeds $2D_e$, no correction factor need be applied.

Notes:

1. The factors in the table are applicable to groups of cables all of one size. The value of current derived from application of the appropriate factors is the maximum continuous current to be carried by any of the cables in the group.

2. If, due to known operating conditions, a cable is expected to carry not more than 30% of its grouped rating, it may be ignored for the purpose of obtaining the rating factor for the rest of the group.

For example, a group of N loaded cables would normally require a group reduction factor of C_g applied to the tabulated I_t. However, if M cables in the group carry loads which are not greater than $0.3\ C_g I_t$ amperes the other cables can be sized by using the group rating factor corresponding to (N–M) cables.

† Not applicable to MI cables.

Table 3.5A　Conduit factors for runs incorporating bends (Extract from IEE On-site guide Table 5D)

Length of run (m)	Straight				One bend				Two bends				Three bends				Four bends			
	16*	20	25	32	16	20	25	32	16	20	25	32	16	20	25	32	16	20	25	32
1					188	303	543	947	177	286	514	900	158	256	463	818	130	213	388	692
1.5					182	294	528	923	167	270	487	857	143	233	422	750	111	182	333	600
2					177	286	514	900	158	256	463	818	130	213	388	692	97	159	292	529
2.5					171	278	500	878	150	244	442	783	120	196	358	643	86	141	260	474
3					167	270	487	857	143	233	422	750	111	182	333	600				
3.5	179	290	521	911	162	263	475	837	136	222	404	720	103	169	311	563				
4	177	286	514	900	158	256	463	818	130	213	388	692	97	159	292	529				
4.5	174	282	507	889	154	250	452	800	125	204	373	667	91	149	275	500				
5	171	278	500	878	150	244	442	783	120	196	358	643	86	141	260	474				
6	167	270	487	857	143	233	422	750	111	182	333	600								
7	162	263	475	837	136	222	404	720	103	169	311	563								
8	158	256	463	818	130	213	388	692	97	159	292	529								
9	154	250	452	800	125	204	373	667	91	149	275	500								
10	150	244	442	783	120	196	358	643	86	141	260	474								

* Diameter of conduit in millimetres.

Table 3.5B   *Cable factors for long straight runs, or runs incorporating bends in conduit*

Type of conductor	Conductor, cross-sectional area (mm²)	Factor
Solid or stranded	1	16
	1.5	22
	2.5	30
	4	43
	6	58
	10	105

Table 3.6   *Single-core PVC-insulated cables in trunking*

For each cable it is intended to use, obtain the appropriate factor from Table A.

Add all the cable factors so obtained and compare with the factors for trunking given in Table B.

The size of trunking which will satisfactorily accommodate the cables is that size having a factor equal to or exceeding the sum of the cable factors.

Table A			Table B	
Cable factors for trunking			Factor for trunking	
Type of conductor	Conductor cross-sectional area (mm²)	Factor	Dimensions of trunking (mm × mm)	Factor
Solid	1.5	7.1	50 × 37.5	767
	2.5	10.2	50 × 50	1037
Stranded	1.5	8.1	75 × 25	738
	2.5	11.4	75 × 37.5	1146
	4	15.2	75 × 50	1555
	6	22.9	75 × 75	2371
	10	36.3	100 × 25	993
			100 × 37.5	1542
			100 × 50	2091
			100 × 75	3189
			100 × 100	4252

For sizes and types of cables and sizes of trunking other than those given above, the number of cables installed should be such that the resulting space factor does not exceed 45%.

Table 3.7   *Cable resistance for solid copper and aluminium conductors*

Cross-sectional area (mm^2)	Resistance (ohms per 1000 metres) Bare conductors at 20°C Copper	Aluminium
0.5	36.00	
1.0	18.10	
1.5	12.10	18.10
2.5	7.41	12.10
4.0	4.61	7.41
6.0	3.08	4.61
10.0	1.83	3.08
16.0	1.15	1.91
25.0	0.727	1.20
35.0	0.524	0.868

Note: For live conductor resistance under short circuit fault conditions, the values given above must be multiplied by the following factors:

For PVC insulation	1.38
For 85°C rubber insulation	1.53
For Mineral Insulation	1.55

For protective conductor resistance under short circuit fault conditions, the values given above must be multiplied by the following factors:

For PVC insulation	1.30
For 85°C rubber insulation	1.42.

4 *Distribution of supplies in buildings*

If a supply is to be obtained from an electricity supplier, then the first step is to get in touch with the local engineer. He should be approached and given full details of the proposed installation before any work is commenced, as it is possible that the load cannot be accepted without an extension or reinforcement of their supply cables, and the consumer may have to contribute to the cost of any such extensions.

The electricity supplier will also be able to advise as to the most suitable tariff for the consumer, and also the position of the point of entry of their supply cable. In large industrial installations they will probably bring in an 11 kV supply, and it will be necessary for the consumer to provide space for switchgear and an 11 kV to 415/240 V transformer on the premises (Fig. 4.1).

For small domestic installations every effort should be made to ensure that the intake position (Fig. 4.2) is in a convenient position, and not in the corridor near the front door, or under a staircase, as used to be the practice of many area boards in the past in order to minimise the length of service cable.

Metering cabinets are available which can be built into outside walls with access from front and back, and the meters can be read from the outside.

If the electricity supplier agrees to give a supply, they must in compliance with Electricity Safety, Quality and Continuity Regulation 27(1) 'notify the consumer the declared voltage, the number of phases and frequency of the supply'. Regulation 27(2) states that they must also constantly maintain the type of current, frequency and the declared voltage, except that the frequency may be varied by 1 per cent, and the voltage not exceeding 10 per cent above or 6 per cent below the declared voltage.

After these initial details have been arranged, and the appropriate tariff agreed, the electricity distributor will install their service cable and cutouts. These service cables will usually be underground, but

Fig. 4.1   *A factory substation. Incoming supplies are from an 11 kV ring main and feed two 800 kVA, 11 kV to 415 V transformers. The distribution board in the substation use 4-pole 1600 A Air Circuit Breakers (ACBs) set at 1200 A and a 4-pole bus coupler rated at 2000 A (W.T. Parker Ltd)*

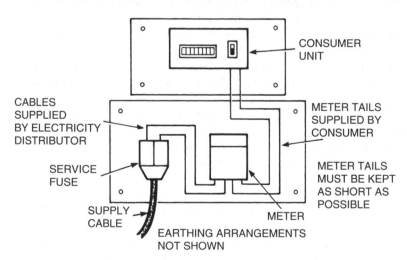

Fig. 4.2   *Typical domestic intake arrangement*

in some areas they may be overhead. Before any internal wiring is commenced it is important that the exact position of the intake is agreed. In Great Britain the system of supply is normally 415/240 V a.c., 50 Hz, but in countries where the supply is d.c. the system is usually 3-wire

Fig. 4.3 *Typical layout of switch-gear for large commercial premises. Note the main switches for normal and standby supplies (left), subsection switches (lower) and distribution boards (right) (Square D Ltd)*

Fig. 4.4 *Main switchboard with mains and generator circuit breakers, and bus section switches with outgoing MCCBs (Durham Switchgear Ltd)*

200/400 V, the lower voltage being between one outer conductor and the middle wire and the higher voltage between the two outers.

For domestic premises it is always advisable not to introduce more than one phase of a supply, unless 3-phase equipment is likely to be used.

Where more than one phase is introduced special precautions have to be taken to avoid shock at above 250 V, and these precautions are covered by the IEE Regulations, and are mentioned on page 127 of this book.

Main switchgear

Every installation, of whatever size, must be controlled by one or more main switches. IEE Regulation 476-01 requires that every installation shall be provided with a means of isolation. A linked switch or circuit breaker at the origin shall switch the following conductors of the incoming supply:

(1) Both live conductors when the supply is single-phase a.c.
(2) All poles of a d.c. supply.
(3) All phase conductors in a T.P., or TP&N., TN-S or TN-C-S system supply.
(4) All live conductors in a T.P., or TP&N., TT, or IT system supply.

This must be readily accessible to the consumer and as near as possible to the supply cutouts. The Electricity at Work Regulations 1989 says that 'suitable means ... shall be available for ... cutting off the supply of electrical energy to any electrical equipment'. The type and size of main switchgear will depend upon the type and size of the installation and its total maximum load. Every detached building must have its own means of isolation.

Cables from the supply cutout and the meter to the incoming terminals of the main switch must be provided by the consumer, they should be kept as short as possible, must not exceed 3 m, and must be suitably protected against mechanical damage. These cables must have a current rating not less than that of the service fuse.

The electricity distributor should be consulted as to their exact requirements as they may vary from district to district.

Whatever size of switchgear is installed to control outgoing circuits, the rating of the fuses or the setting of the circuit breaker overloads must be arranged to protect the cable which is connected for the time being. If a distribution circuit cable is rated to carry 100 A then the setting of the excess-current device must not exceed 100 A.

IEE Regulation 130-03 says that every circuit must be protected against overcurrent by a device which will operate automatically and is of adequate breaking capacity. The protective device may, therefore, serve two functions, first to prevent overloading of the

Fig. 4.5 *Main distribution board installed in industrial premises (Square D Ltd)*

circuit, secondly to be capable of interrupting the circuit rapidly and without danger when a short circuit occurs.

Although protective devices must be capable of opening the circuit almost instantaneously in the event of a short circuit, they must be sufficiently discriminating so as not to operate in the event of a temporary overload.

Rewirable fuses The rewirable fuse made to BS 3036 is now mainly confined to domestic installations, for the reasons given in Chapter 2.

High breaking capacity fuses The HBC fuse has been designed to clear a potentially dangerous short circuit within 0.0013 s. It would be a great disadvantage if these same fuses operated with equal promptitude every time a temporary overload occurs, and would result in considerable expense for fuselink replacements.

For this reason HBC fuselinks are designed so that they will withstand as much as 5 times full load current for a few seconds, by which time the fault will probably be cleared by a final circuit protective device (Fig. 4.18), or local control gear. If main HBC fuses are carefully selected and graded so as to function with discrimination, the final circuit protective device will take care of all normal overloads and short circuits, and the main HBC fuse will operate only when the short circuit is in the feeder cable the fuse is protecting, or in the event of the cumulative load of the final circuits exceeding the rating of the main fuses.

Circuit breakers   Modern circuit breakers are designed to handle safely heavy short-circuit currents in the same manner as HBC fuses.

Such circuit breakers have a number of advantages over other types of circuit protection. However care is needed in selection and maintenance to ensure compliance with Regulation 5 of the Electricity at Work Regulations, 1989, which requires that the arrangements must not give rise to danger, even under overload conditions. If a moulded case circuit breaker has had to clear faults at its full rated breaking capacity, it may need to be replaced to ensure that it can interrupt a fault current safely.

Circuit breakers do have some inherent advantages. In the event of a fault, or overload, all poles are simultaneously disconnected from the supply. Some types of device are capable of remote operation, for example by emergency stop buttons, and some have overloads capable of adjustment within pre-determined limits.

Selection of switchgear of suitable capacity

In Great Britain the electricity distributors offer alternative tariffs, and they will always advise consumers as to which is the most favourable tariff after taking into account various factors, such as installed load, type of load, estimated maximum demand and so on.

For large industrial installations it may be an advantage for a consumer to purchase electricity at high-voltage, although this will entail capital expenditure for h.v. switchgear and transformers.

Whatever type of installation, whether domestic, commercial, or industrial, it is necessary to consult the electricity distributor at an early stage in the design of an installation.

As has already been pointed out, the main rule which governs all installation work is 'that all apparatus must be sufficient in size and power for the work they are called upon to do'. This applies especially to main switchgear, and it is important to ensure that it is in no danger of being overloaded.

To determine the size required it is necessary to add up the total connected lighting, heating, power and other loads, and then calculate the total maximum current which is likely to flow in the installation. This will depend upon the type of installation, how the premises will be used, whether there are alternative or supplementary means of heating and cooking, and other considerations such as diversity described in Chapter 2 and below.

IEE Regulation 311-01-01 states that in determining the maximum demand of an installation or parts thereof, diversity may be taken in account.

Table 4.1 gives guidance on diversity, but it is emphasised that the calculation of diversity is a matter calling for special knowledge and experience.

By consulting this table a reasonable estimate can be obtained as to what the maximum load is likely to be, but it must be stressed that each installation must be dealt with on its own merits.

To take an example of a domestic installation with a single tariff:

Item In Table 4.1	Connected load		Expected maximum demand	
1	Installed lighting	10 A	66% of installed load	= 6.6 A
2	Installed fixed heating	30 A	100% of first 10 A plus 50% of excess of 10 A	= 20 A
9	Installed general purpose socket outlets (Two ring circuits)	64 A	100% of current demand of largest circuit (32 A) plus 40% current demand of other circuits (12.8 A)	= 44.8 A
3	Installed cooker	45 A	10 A plus 30% f.l. of remaining connected appliances plus 5 A for socket in unit	= 22 A
	Total	149 A		93.4 A

Note: The load on socket-outlet circuits is taken as the rating of the protective device.

In this case a 100 A main switch should be provided, unless it is anticipated to increase the load considerably in the foreseeable future, in which case a larger switchfuse should be installed.

A different example would be for a church where electric lighting and heating is installed; here it would be most likely that the whole load will be switched on at one time and therefore the main switchgear must be suitable for the total installed load.

Table 4.1 Table of typical allowances for diversity (IEE On-site guide, Table 1B)

Purpose of final circuit fed from conductors or switchgear to which diversity applies	Type of premises		
	Individual household installations, including individual dwellings of a block	Small shops, stores, offices and business premises	Small hotels, boarding houses, guest houses, etc.
1 Lighting	66% of total demand	90% of total current demand	75% of total current demand
2 Heating and power (but see 3–8 below)	100% of total current demand up to 10 A +50% of any current demand in excess of 10 A	100% f.l. of largest appliance +75% of remaining appliances	100% f.l. of largest appliance +80% f.l. of second largest appliance +60% of remaining appliances
3 Cooking appliances	10 A +30% f.l. of connected cooking appliances in excess of 10 A +5 A if socket-outlet incorporated in unit	100% f.l. of largest appliance +80% f.l. of second largest appliance +60% f.l. of remaining appliances	100% f.l. of largest appliance +80% f.l. of second largest appliance +60% f.l. of remaining appliances
4 Motors (other than lift motors which are subject to special consideration)		100% f.l. of largest motor +80% f.l. of second largest motor +60% f.l. of remaining motors	100% f.l. of largest motor +50% f.l. of remaining motors
5 Water heaters (instantaneous type)*	100% f.l. of largest appliance +100% of second largest appliance +25% f.l. of remaining appliance	100% f.l. of largest appliance +100% of second largest appliance +25% f.l. of remaining appliances	+100% f.l. of largest appliance +100% of second largest appliance +25% f.l. of remaining appliances

6 Water heaters (thermostatically controlled)	NO DIVERSITY ALLOWABLE†		
7 Floor warming installations	NO DIVERSITY ALLOWABLE†		
8 Thermal storage space heating installations	NO DIVERSITY ALLOWABLE†		
9 Standard arrangements of final circuits in accordance with IEE Appendix 5	100% of current demand of largest circuit +40% of current demand of every other circuit	100% of current demand of largest circuit +50% of current demand of every other circuit	
10 Socket outlets other than those included in 9 above and stationary equipment other than those listed above	100% of current demand of largest point of utilisation +40% of current demand of every other point of utilisation	100% of current demand of largest point of utilisation +75% of current demand of every other point of utilisation	100% of current demand of largest point of utilisation +75% of current demand of every point in main rooms (dining rooms, etc.) +40% of current demand of every other point of utilisation

* For the purpose of this table an instantaneous water heater is deemed to be a water heater of any loading which heats water only while the tap is turned on and therefore uses electricity intermittently.

† It is important to ensure that the distribution boards are of sufficient rating to take the total load connected to them without the application of any diversity.

Main switchgear for domestic installations

It is usual to install a domestic consumer unit as the main switchgear, and also as the distribution point in a small or domestic installation. A wide range of makes and types of consumer unit is available, and some of these are shown in the accompanying illustrations. These units usually consist of a main switch of up to 100 A capacity, and an associated group of up to 16 single-pole fuseways for overcurrent protection of individual circuits. No main fuse is normally used with these units as the supply undertaking's service fuse will often provide the necessary protection of the tails connecting the fuse to the consumer unit. To ensure that this is so, a knowledge of the prospective short circuit currents is necessary, and the breaking capacity of the devices to be used. This is covered in more detail in Chapter 2 of this book.

Protective devices fitted in the unit can be semi-enclosed fuses, HBC fuses or miniature circuit breakers. The relative advantages and disadvantages of each of these are discussed in Chapter 2, and as explained the use of semi-enclosed fuses could well result in larger cables being required for circuits protected against overload current.

Consumer units are also available which incorporate residual current circuit breakers as part of the unit. These may be arranged to feed all the ways in the unit, or to feed only a proportion of them.

Fig. 4.6   *A cut-away view of a moulded case circuit breaker (MCCB). This device incorporates both bimetallic and magnetic trip mechanisms to open the contacts under overload or short circuit conditions. The operating toggle has three positions and shows when the breaker has tripped. A range of auxiliary components can be fitted such as undervoltage releases, or control interlocks. These MCCBs can be obtained with breaking capacities up to 150 kA (Merlin Gerin Ltd)*

The former is especially useful where a TT system is in use, as the residual current protection enables the regulations for protection against indirect contact to be complied with. It should be noted that there is not necessarily any benefit in providing residual current protection on circuits where it is not strictly necessary. Provided the installation design is such that the correct disconnection times are obtainable, normal overcurrent protection will suffice. To take an example, a residual current device is needed for any sockets intended for equipment being used outdoors. If this residual current device is one in a consumer unit which acts on all the circuits, a fault on one circuit will trip the residual current circuit breaker and disconnect the whole installation. In order to avoid any inconvenience to the users, it would be better therefore to provide the residual current protection only on the circuits which demand it.

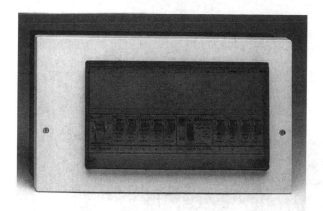

Fig. 4.7   *A 9-way consumer unit with MCB and RCCD protection (Crabtree Electrical Industries Ltd)*

Fig. 4.8   *Seven-way metal-clad consumer unit with miniature circuit breaker protection and incorporating a residual current circuit breaker (MEM Ltd)*

Fig. 4.9   *Consumer unit fitted with circuit breakers and a main residual current circuit breaker (MK Ltd)*

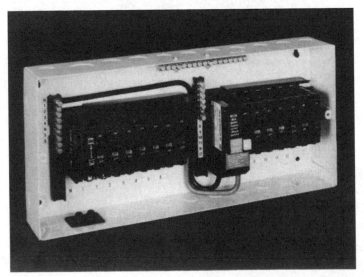

Fig. 4.10   *Twelve-way consumer unit with front removed to show the 100 A main switch and six of the twelve ways protected by a 30 mA residual current circuit breaker (Square D Ltd)*

Fig. 4.11 *The MCCB Panelboard is available in a range of ratings up to 800 A and with up to 18 outgoing MCCBs fitted. The version illustrated incorporates a metering pack at the top (Bill Switchgear Ltd)*

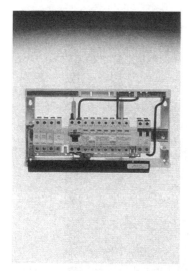

Fig. 4.12 *Domestic consumer unit with the cover removed showing the main switch, MCBs, a residual current device and fuse corners (MEM Ltd)*

Main switchgear for industrial installations

Main switchgear for industrial and other similar installations, such as commercial buildings, hospitals and schools, will be designed and rated according to the maximum current that is likely to be used at

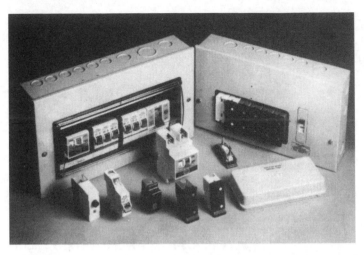

Fig. 4.13   *Comparison of two types of six-way consumer unit, and also a selection of the protective devices which can be fitted. The consumer unit on the left has miniature circuit breaker and residual current protection, and that on the right incorporates semi-enclosed fuses (George H. Scholes PLC)*

peak periods, and in extreme cases (such as churches) might be as much as 100% of the installed load. For such installations it is usual to provide main switchgear, not only of sufficient size to carry the installed load, but to allow ample margins for future extensions to the load.

Large industrial and commercial installations   For loads exceeding 200 kVA it is usual for one or more h.v. transformers to be installed on the consumer's premises. The electricity supplier should be consulted at an early stage to ascertain whether space for a substation will be required, and to agree on its position. It is important that it should be sited as near as possible to the heaviest loads so as to avoid long runs of expensive low-voltage cables.

If heavy currents have to be carried for long distances the size of the cables would have to be increased to avoid excessive voltage drop. This not only increases the cost of the cables, but there would be power losses in the cables for which the consumer will have to pay. It might therefore be advisable to put the substation in the middle of a factory building. 'Package' substations are available which take up a minimum of space. If space is very restricted then the substation could be erected on a platform at high level providing sufficient working space, access and egress are provided.

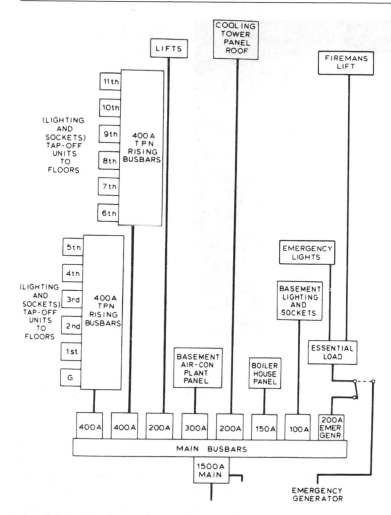

Fig. 4.14   *Distribution diagram for typical commercial multistorey building – meters on each floor if required*

When installing l.v. (low voltage) switchboards for large installations where the supply is derived from a local h.v. transformer, due consideration must be given to the potential fault current which could develop in the event of a short circuit in or near the switchboard. For example, a 1000 kVA 11 kV/415 V 3-phase transformer would probably have a reactance of 5%, and therefore the short-circuit power at the switchboard could be as much as 20 MVA.

The greater the impedance of the cables from the secondary of the transformer to the l.v. switchboard, the less will be the potential

Fig. 4.15   *A switchboard for use in a commercial premises. The board incorporates 12 outgoing switches as well as main Air Circuit Breakers (Pandelco Ltd)*

short-circuit current, and therefore these cables should not be larger than necessary. Most l.v. switchboards are designed to clear faults up to 32 MVA and would therefore be quite capable of clearing any short circuit current imposed upon a 1000 kVA transformer.

If, however, two 1000 kVA transformers were connected in parallel, then the potential fault current could be as much as 40 MVA. As this exceeds the rupturing capacity of standard switchboards it would entail the installation of a much more expensive switchboard, or special high-reactance transformers. It is usual not to connect transformers having a combined rating exceeding 1500 kVA to a standard medium voltage switchboard having a fault rating of 32 MVA. For higher combined ratings it is usual to split the l.v. switchboard into two or more separate sections, each section being fed from a transformer not exceeding 1500 kVA.

Interlocked bus-section switches should be provided to enable one or more sections of the switchboard to be connected to any one transformer in the event of one transformer being out of action, or under circumstances when the load on the two sections of the switchboard is within the capacity of one transformer. Fig. 4.16 shows such an arrangement. Always make sure that the transformer remaining in service is capable of carrying the load, otherwise it could become

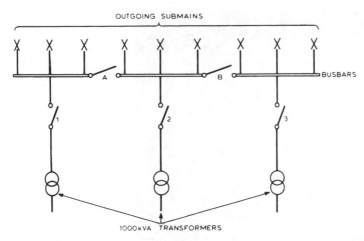

Fig. 4.16 *Arrangement of bus-section switches on low-voltage switchboard (single line diagram) 1, 2 and 3: main switches. A and B: bus-section switches. Bus-section switches A and B are normally open. These are interlocked with main switches 1, 2 and 3. A can only be closed when 1 or 2 are in Off position. B can only be closed when 2 or 3 are in Off position. This enables one transformer to take the load of two sections of the l.v. switchboard if required*

overloaded. It may be necessary to switch off non-essential loads before closing a bus-section switch.

Open-type switchboards An open-type switchboard is one which has exposed current-carrying parts on the front of the switchboard. This type of switchboard is rarely used nowadays but where these exist a handrail or barrier has to be provided to prevent unintentional or accidental contact with exposed live parts. They must be located in a special switchroom or enclosure and only a competent person may have access to these switchboards.

Protected-type switchboards A protected type switchboard (Fig. 4.17) is one where all of the conductors are protected by metal or other enclosures. They may consist of a metal cubicle panel, or an iron frame upon which is mounted metal-clad switchgear. They usually consist of a main switch, busbars and circuit breakers or fuses controlling outgoing circuits. Circuits and conductors are normally segregated within the switchboard and various levels of segregation may be used.

Electricity at Work Regulation 15 gives other requirements which apply equally to open-type and protected-type switchboards. These include such matters as the need for adequate space behind and in front of switchboards; there shall be an even floor free from obstructions, all

Fig. 4.17   *A protected switchboard with separate lockable compartments to house the incoming and outgoing cabling. Although these can be accessed from the front of the panel, it is still essential to allow space behind the panel to allow subsequent maintenance to be carried out at the rear (Pandelco Ltd)*

parts which have to be handled shall be readily accessible, it must be possible to trace every conductor and to distinguish between these and those of other systems, and all bare conductors must be placed or protected so as to prevent accidental short circuit.

Busbar chambers   Busbar chambers which feed two or more circuits must be controlled by a switch, circuit breaker, links or fuses to enable them to be disconnected from the supply to comply with IEE Regulation 131-14-01 (Fig. 4.21).

Earthed neutrals   To comply with IEE Regulation 131-13-02, and Regulation 9 of the Electricity at Work Regulations 1989, no fuse or circuit breaker other than a linked circuit breaker shall be inserted in

Fig. 4.18 *Heavy current air circuit breaker incorporates electronic control units offering a range of characteristics and options including data communications and inter-tripping (Merlin Gerin Ltd)*

Fig. 4.19 *Triple-pole switch fuse fitted with HRC fuses (Bill Switchgear Ltd)*

an earthed neutral conductor, and any linked circuit breaker inserted in an earthed neutral conductor shall be arranged to break all the related phase conductors.

These regulations cover PME supplies and the above rule applies throughout the installation, including 2-wire final circuits. This means that no fuses may be inserted in the neutral or common return

Fig. 4.20   *Substation packages are available which can speed up installation on site. This 11 kV to 433 V substation, rated at 1000 kVA is shown above ready for wiring. The lower illustration shows the cladding installed (GEC Transformers Ltd)*

wire, and the neutral should consist of a bolted solid link, or part of a linked switch which completely disconnects the whole system from the supply. This linked switch must be arranged so that the neutral makes before, and breaks after the phases.

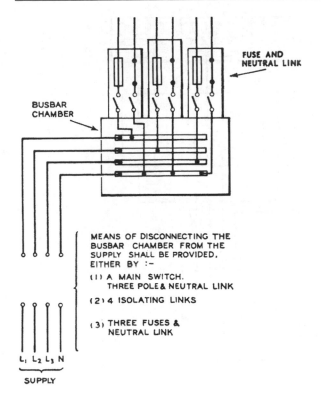

Fig. 4.21 *Isolation of busbar chamber. Busbar chambers must have a means of disconnection from the supply*

Distribution boards

A distribution board may be defined as 'a unit comprising one or more protective devices against overcurrent and ensuring the distribution of electrical energy to the circuits'. It is wise to select distribution boards which provide plenty of wiring space and with terminals of adequate size to accommodate the cables which will be connected to them.

Very often it is necessary to install a cable which is larger than would normally be required, in order to limit voltage drop, and sometimes the main terminals are not of sufficient size to accommodate these larger cables. Therefore distribution boards should be selected with main terminals of sufficient size for these larger cables.

Types of distribution boards
There are three types of distribution boards, (1) those fitted with rewirable fuselinks, (2) those fitted with HBC fuselinks, (3) those fitted with circuit breakers.

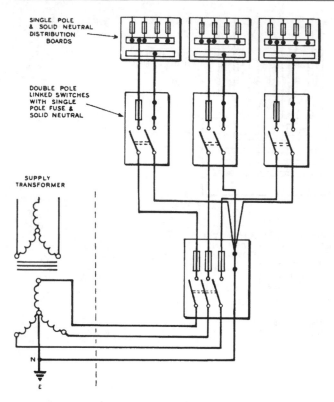

Fig. 4.22 *Single-pole fusing. No switch or circuit breaker (except where linked) or fuse, shall be inserted in an earthed neutral conductor. Where a linked switch or circuit breaker is inserted in an earthed neutral then it shall be arranged to disconnect all the related phase conductors (IEE Regulation 131-13-02)*

As already explained there are many objections to the use of rewirable fuses, as there is very little guarantee that the fuselink chosen at the time of the installation will not be replaced by a larger fuselink. There is much to be said for choosing distribution boards fitted with MCBs or HRC fuselinks. Distribution boards fitted with miniature circuit breakers are more expensive in their first cost, but they have much to commend them, especially as they can incorporate an earth-leakage trip. Miniature circuit breakers are obtainable in ratings from 5 A to 60 A, all of which are of the same physical size, and are therefore easily interchangeable. (They must not of course be inter-changed without first making sure that the cables they protect are of the correct rating.)

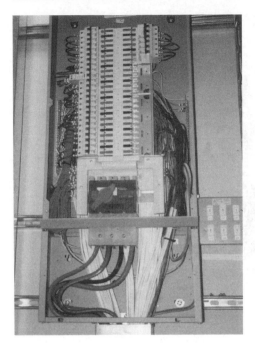

Fig. 4.23 *MCB distribution boards form a convenient way of arranging distribution of supplies. They can be obtained in a range of sizes, and the illustration shows the board with covers removed (W.T. Parker Ltd)*

Every distribution board must be connected to either a main switch fuse or a separate way on a main distribution board. Every final circuit must be connected to either a switch fuse, or to one way of a distribution board.

Positions of distribution boards

Distribution boards should preferably be sited as near as possible to the centre of the loads they are intended to control. This will minimise the length and cost of final circuit cables, but this must be balanced against the cost of submain cables.

Other factors which will help to decide the best position of distribution boards are the availability of suitable stanchions or walls, the ease with which circuit wiring can be run to the position chosen, accessibility for replacement of fuselinks, and freedom from dampness and adverse conditions. (If exposed to the weather or damp conditions, a distribution board must be of the weather-proof type.)

If a distribution board is recessed into a wall which is constructed of combustible material such as wood, the case must be made of metal or other non-combustible material.

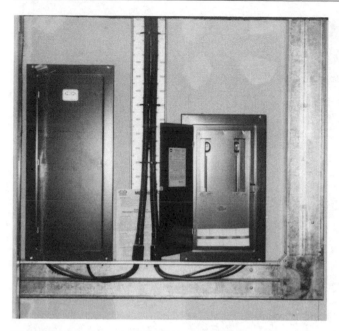

Fig. 4.24   MCB distribution boards in the process of being installed. The incoming cable is run on cable tray, and trunking for the outgoing cables is in place. The individual MCBs have not yet been fitted (William Steward & Co. Ltd)

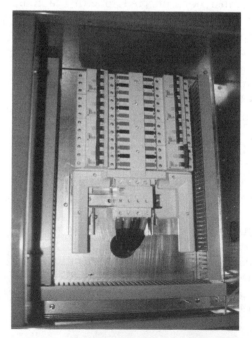

Fig. 4.25   Another type of MCB distribution board. The board illustrated is fitted with 8 linked 3-phase MCBs, fed by a 3-phase and neutral cable from an isolator in a separate compartment (Pandelco Ltd)

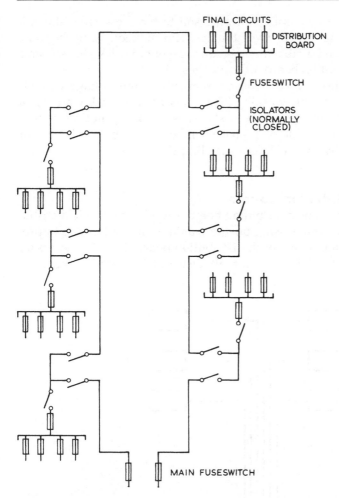

FINAL CIRCUITS

DISTRIBUTION BOARD

FUSESWITCH

ISOLATORS (NORMALLY CLOSED)

MAIN FUSESWITCH

Fig. 4.26 *Single line diagram of a typical ring main feeding six distribution boards. The cables feeding the ring will share the load and may therefore be reduced accordingly. This arrangement enables the ring to be broken by one of the isolators in the event of a fault at one end of the ring, in which case the load must be reduced*

Supplies exceeding 230 V a.c.

Where distribution boards (which are fed from a supply exceeding 230 V) feed circuits with a voltage not exceeding 230 V then precautions must be taken to avoid accidental shock at the higher voltage between the terminals of two lower voltage boards.

For example, if one distribution board were fed from the brown phase of a 415/240 V system of supply, and another from the grey phase, it would be possible for a person to receive a 415 V shock if live parts of both boards were touched simultaneously. In the same way it would be possible for a person to receive a 415 V shock from a 3-phase distribution board, or switchgear.

IEE Regulation 514-10 requires that where the voltage exceeds 230 V, a clearly visible warning label must be provided, warning of the maximum voltage which exists. These warning notices should be fixed on the outside of busbar chambers, distribution boards or switchgear, whenever voltage exceeding 230 V exists.

Feeding distribution boards

When more than one distribution board is fed from a single distribution circuit, or from a rising busbar trunking, it is advisable to provide local isolation near each distribution board. It is also necessary to provide a local isolator for all distribution boards which are

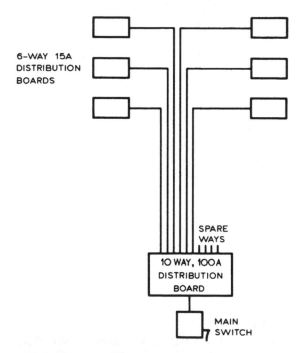

Fig. 4.27 *Single line diagram showing six final distribution boards fed by radial distribution circuits from a main distribution board. It is recommended that distribution boards located remote from main switchgear be provided with local isolators*

situated remote from the main switchboard, since IEE Regulations 131-14-01 and 476-01-01 call for every installation and circuit to be provided with isolation and switching.

If the main or submain cables consist of bare or insulated conductors in metal trunking, it is very often convenient to fit the distribution boards adjacent to the rising trunking, and to control each board with fusible cutouts or a switchfuse.

Circuit charts and labelling

IEE Regulation 514-09-01 requires that diagrams, charts or tables shall be provided to indicate for each circuit:

(1) the type and composition of each circuit
(2) the method of protection against indirect contact
(3) identification, location of isolators and protective devices.

These should be fixed in, or in the vicinity of, the distribution board, and fitted in glazed frames or in plastic envelopes for protection.

Marking distribution boards

All distribution boards should be marked with a letter or number, or both, preferably with the prefix L for lighting, S for sockets, and P for power.

They should also be marked with the voltage and the type of supply, and if the supply exceeds 250 V a DANGER notice must be fixed.

When planning an installation a margin of spare fuseways should be provided – usually about 20% of the total and this must be matched by an increase in the current-carrying capacity of the distribution cables.

Metal distribution boards should be provided with plugged holes to enable additional conduits or multicore cables to be easily connected in future.

Colour identification of cables and conductors

IEE Regulation 514 lays down the requirement for identification of conductors. The 2004 amendment to the IEE Wiring Regulations (BS 7671) introduces a new standard of wiring colours for fixed wiring, bringing harmonisation across Europe. The change is a significant one for the UK and the changeover timescale from old to new wiring colours is laid down in the foreword to the Regulations. Installations commencing on site after 31st March 2006 are to comply with the marking requirements of section 514 as amended, cores being identified using the new harmonised colours. Installations commencing on site before that date may be installed to section 514 as amended or to the earlier standard, that is to say they may use the old or the new (harmonised) colours, but NOT both. Unambiguous marking is essential and the new IEE Regulations appendix 7 gives guidance on this matter.

IEE Regulation 514-03-02 states that every core of a cable shall be identifiable at its terminations and preferably throughout its length and IEE Table 51 specifies the alphanumeric and colour identification to be used. There are a few exceptions to this and these include concentric conductors, metal sheaths or armouring when used as a protective conductor and bare conductors where permanent identification is not practicable.

Table 4.2 summarises the colour requirements and includes extracts from IEE Table 51. The new colours will not be totally unfamiliar, as brown and blue have been in use for flexible cables for some time. However, the introduction of new colours for fixed wiring including phase conductors on three-phase circuits means that clear identification is essential to prevent confusion.

This is particularly so at the interfaces which will occur where wiring alterations or additions are made to existing installations. Although colour identification alone is permitted at interfaces in single-phase installations, additional permanent alphanumeric marking is required in two- or three-phase schemes. It will be appreciated that the old phase colour blue must not be confused with the new neutral cable colour which is also blue. The table lays down alphanumeric symbols to be used and, at a three-phase interface, both existing and additional cores shall be marked 'N' for neutral conductors and 'L1', 'L2' or 'L3' for phase conductors.

In any installation, whether single- or three-phase, where two different colour standards are present, a warning notice must be affixed at or near distribution boards worded:

CAUTION

This installation has wiring colours to two versions of BS7671. Great care should be taken before undertaking extension, alteration or repair that all conductors are correctly identified.

Switch wires

It is usual to run a two-core and cpc cable with cores coloured brown and blue to a switch position, both conductors being phase conductors. In such a case, the blue conductor must be sleeved brown or marked 'L'

Table 4.2 *Colour identification of conductors and cables (includes extracts from IEE Table 51)*

Function	Alpha-numeric	Colour (IEE Table 51)	Old Fixed Wiring Colour (see text)
Protective conductors		Green and yellow	Green and yellow
Functional earthing conductor		Cream	Cream
a.c. power circuit (including lighting)			
Phase of single phase circuit	L	Brown	Red
Phase 1 of three phase circuit	L1	Brown	Red
Phase 2 of three phase circuit	L2	Black	Yellow
Phase 3 of three phase circuit	L3	Grey	Blue
Neutral for single or three phase circuit	N	Blue	Black
Two wire unearthed d.c. circuits			
Positive	L+	Brown	Red
Negative	L−	Grey	Black
Two wire earthed d.c. circuit			
Positive (of negative earthed) circuit	L+	Brown	Red
Negative (of negative earthed) circuit	M	Blue	Black
Positive (of positive earthed) circuit	M	Blue	Black
Negative (of positive earthed) circuit	L−	Grey	Blue
Three wire d.c. circuit			
Outer positive of two wire circuit derived from three wire system	L+	Brown	Red
Outer negative of two wire circuit derived from three wire system	L−	Grey	Red
Positive of three wire circuit	L+	Brown	Red
Mid wire of three wire circuit	M	Blue	Black
Negative of three wire circuit	L−	Grey	Blue
Control circuits, extra low voltage etc.			
Phase conductor	L	Brown, Black, Red, Orange, Yellow, Violet, Grey, White, Pink or Turquoise	
Neutral or mid wire	N or M	Blue	

at the terminations. The same applies to the black and grey cores of three-core cables if used in intermediate or two-way switched circuits.

MI cables

At the termination of these cables, sleeves or markers shall be fitted so that the cores are identified and comply with IEE table 51.

Bare conductors

Where practical, as in the case of busbars, these are to be fitted with sleeves, discs, tapes, or painted to comply with IEE table 51. An exception is made where this would be impractical such as with the sliding contact conductors of gantry cranes, but even there, identification would be possible at the terminations.

PEN conductor

A PEN (combined protective and neutral) conductor may be coloured either green and yellow throughout its length with, in addition, blue marking at the terminations or blue throughout its length with, in addition, green and yellow marking at the terminations.

Motor circuits

When wiring to motors, the colours specified in IEE table 51 should be used right up to the motor terminal box. For slip-ring motors, the colours for the rotor cables should be the same as those for phase cables, or could be all one colour except blue, green, or green and yellow.

For star delta connections between the starter and the motor, use Brown for A1 and A0, Black for B1 and B0 and Grey for C1 and C0. The 1 cables should be marked to distinguish them from the 0 cables.

Conduit

The identification colour is orange and should be used whenever it is desirable to distinguish electrical services from others such as gas, oil or air pipes.

Distribution circuits

Distribution circuits (sometimes referred to as Submains) are those which connect between a main switchboard, a switch fuse, or a main distribution board to subdistribution boards. The size of these cables will be determined by the total connected load which they supply, with due consideration for diversity and voltage drop, and the other factors described in Chapter 2.

Distribution circuits may be arranged to feed more than one distribution board if desired. They may be arranged to form a ring circuit, or a radial circuit looping from one distribution board to another. Where a distribution circuit feeds more than one distribution board its size must not be reduced when feeding the second or subsequent board, because the cable must have a current rating not less than the fuse or circuit breaker protecting the submain (IEE Regulation 433-02-01).

If a fuse or circuit breaker is inserted at the point where a reduction in the size of the cable is proposed, then a reduced size of cable may be used, providing that the protective device is rated to protect the cable it controls.

Protective multiple earthing (PME)

What the PME type of supply amounts to is using the protective conductor as a combined earth/neutral conductor. It is sometimes used where there is overhead distribution, and where it is difficult to obtain a sufficiently low earth resistance from the supply transformer to the consumer's terminal. In such a case the neutral conductor is also the earth conductor and it is bonded to earth, not only at the transformer position, but also at the consumer's terminal position. The Conditions of Approval for this system contain very stringent requirements.

The wiring for consumers' installations, including distribution and circuit wiring may (if approved) be carried out on the PME system. Some of the requirements for consumers' installations are as follows:

(1) The electricity distributor shall be consulted to determine any special requirements concerning the size of protective conductors.

(2) All precautions must be taken to avoid the possibility of an open circuit in the neutral conductor.

(3) Bonding leads must be connected to the earthing terminals of all metal structures, metal pipes and other metal services that are (or may reasonably be expected to become) in electrical contact with the general mass of earth, and that are so situated that simultaneous contact may reasonably be expected to be made by any person with such structures, pipes or other metalwork on the one hand, and with the exposed non-current-carrying metalwork of the consumer's installation, or any metalwork in electrical contact therewith, on the other hand.

(4) Earth electrodes shall be provided at points not less remote from the transformer than the most remote service line or connection point, and at such other points as will ensure that the resistance to earth in the neutral conductor is satisfactory and the protection system operative. The overall resistance shall not exceed 20 Ω.

It should be noted that regulations now exist which specifically prohibit the use of a PME supply in petrol filling stations. The details are given in Health and Safety booklet HS (G) 41 'Petrol Filling Stations: Construction and Operation'. The reason for this prohibition is to prevent the risk of electrical return currents flowing back to earth through the metallic parts of the underground supply pipes and storage tanks.

Cable manufacturers make special armoured multicore cables for PME supplies. These are available with XLPE (cross-linked polyethylene) insulation. Aluminium conductors and sheath are used, and the cables have a PVC oversheath. The armouring is laid up in such a way that sufficient can be pulled away from the cable without the necessity of cutting it, to enable access to be made to the phase conductors for the purpose of jointing. These special cables are only manufactured in minimum lengths of about 200 m, and it may not be economical to employ PME supplies for distribution cables when only short runs are involved. When the neutral is used as the protective conductor in the supply and in the installation the system becomes TN-C and permission has to be obtained before a TN-C system is installed. See Fig. 2.3.

Power factor

Power factor is an inherent feature in connection with the installation of induction motors. The power factor of an induction motor may be as low as 0.6 which means that only 60 per cent of the current is doing useful work. For average machines a power factor of 0.8 lagging is the general rule.

It is advisable, therefore, to understand what power factor means and how it can be measured and improved. In a practical book of this kind no attempt will be made to give in technical terms the theory of power factor, but perhaps a rudimentary idea can be conveyed.

In an inductive circuit, such as exists in the case of an induction motor, the power in the circuit is equal to the instantaneous value of the voltage multiplied by current in amperes, the product being in watts. A wattmeter, kilowattmeter or kilowatt-hour meter, if placed in circuit, will register these instantaneous values, multiply them, and give a reading in watts, kilowatts or kilowatt-hours.

Actually in an a.c. circuit the voltage, and therefore the current, varies from zero to maximum and maximum to zero with every cycle.

In an inductive a.c. circuit the current lags behind the voltage. For example, if the normal voltage is 400 and the current is 50 A, when the voltage reaches 400 the current may have only reached 30 A and by the time the current has risen to 50 A the voltage would have fallen to 240. In either case the total watts would be 12 000 and not 20 000 as would be the case in a non-inductive circuit. If a separate ammeter and voltmeter were placed in this circuit the voltmeter would give a steady reading of the nominal 400 V and the ammeter would read the nominal 50 A, the product of which would be 20 000 VA.

A watt-meter in the same circuit would, as explained, multiply the instantaneous values of voltage and current and the product in this case would be 12 000 W.

To find the power factor of the circuit we divide watts by volt-amperes, i.e.

$$\frac{12\,000}{20\,000} = 0.6 \text{ power factor}$$

Wattless current

It will be seen that a portion of the current is not doing useful work, and this is called wattless current or reactive power. Although this current is doing no useful work, it is flowing in the distributors' cables and also in the cables throughout the installation.

As already explained, the kilowatt-hour meter does not register this wattless current and, therefore, when current is charged for on the basis of units consumed the distributor is not paid for this current.

They therefore have to either insist that the power factor is improved, or they insert an additional meter which records the maximum kilovolt-amperes used during a given period, and charge a sum based upon this maximum figure.

Consumers who switch on very heavy loads for short periods are penalised, and thus are encouraged to keep their maximum demand down to reasonable limits.

In these circumstances it will be a paying proposition to take steps to improve the power factor, as this will not only reduce the maximum kilovolt-amperes demand considerably, but also unload the cables feeding the motors.

Improving the power factor

This is sometimes accomplished by means of a synchronous motor, but generally it is advisable to install capacitors, which introduce capacitance into the circuit, and which counteract the lagging power factor due to induction.

The capacitor only corrects the power factor between the point at which it is inserted and the supply undertaking's generating plant, and therefore from a commercial point of view so long as it is fitted on the consumer's side of the kilovolt-amperes meter its purpose is served. If, however, it is desired to reduce the current in the consumer's cables, then it is advisable to fit the capacitor as near to the motor as possible.

The capacitor has the effect of bringing the current into step with the voltage, but in that part of the circuit not covered by the capacitor they get out of step again. Therefore in the motor itself the power factor still remains low, and its real efficiency is not improved by the installation of a capacitor.

If maximum demand (MD) is based on kilowatts, any improvement in power factor will not result in a reduction of MD charges; if it is based on kilovolt-amps (kVA), however, a considerable saving in MD charges can often be made.

Apart from any financial savings, the installation of capacitors to improve the power factor will reduce the current in switchgear and cables, and this can be of considerable advantage when these happen to be loaded up to their limits.

5 Design and arrangement of final circuits

The previous chapter dealt with the control and distribution of supply and described the necessary equipment from the supply cutouts to the final distribution boards. The planning and arrangement of final circuits, the number of outlets per circuit, overload protection, the method of determining the correct size of cables, and similar matters are dealt with in this chapter, and it is essential that these matters should be fully understood before proceeding with practical installation work.

Definition of a 'final circuit'

A final circuit is one which is connected to any one way of a distribution board, or a switchfuse feeding one or more outlets, without the intervention of a further distribution board.

An outlet is defined as the termination of fixed wiring feeding a luminaire, socket, or any current consuming appliance. From this it will be seen that a final circuit might consist of a pair of 1 mm² cables feeding a few lights or a very large 3-core cable feeding a large motor direct from a circuit breaker or the main switchboard.

Regulations governing final circuits

IEE Regulation 314-01-04 states 'Where an installation comprises more than one final circuit, each shall be connected to a separate way in a distribution board', and that the wiring to each final circuit shall be electrically separate from that of every other final circuit.

For final circuits the nominal current rating of the fuse or circuit breaker (overcurrent device) and cable will depend on the type of final circuit. Final circuits can be divided into the following types, all of which will need different treatment when planning the size of the conductors and the rating of the overcurrent devices:

Final circuit feeding fixed equipment or 2 A sockets
Final circuit feeding 13 A sockets to BS 1363   (*Continued overpage*)

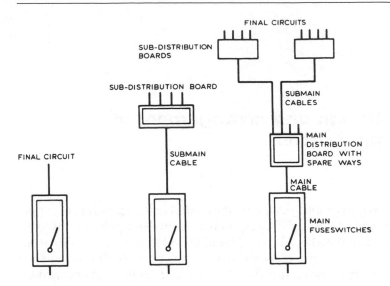

Fig. 5.1   *Single line diagram showing alternate methods of connecting distribution boards*

Table 5.1   *Overcurrent protection of lampholders (IEE Table 55B)*

Type of lampholder as designated in BS EN 60238 and BS EN 61184		Maximum rating of overcurrent protective device protecting the circuit (A)
Bayonet (BS EN 61184)	B15	6
	B22	16
Edison screw (BS EN 60238)	E14	6
	E27	16
	E40	16

Final circuit feeding sockets to BS 196 (5 A, 15 A and 30 A)

Final circuit feeding sockets to BS EN 60309-2 (industrial types 16 A to 125 A)

Final circuit feeding fluorescent or other types of discharge lighting

Final circuit feeding motors

Final circuit feeding cookers.

Final circuit feeding fixed equipment or 2 A sockets

For these final circuits the number of points supplied is limited by their aggregate demand, as determined from Table 5.2. No diversity

Table 5.2 *Suggested assumptions for current demand of equipment*

Point of utilisation of current-using equipment	Current demand to be assumed
(a) Socket outlets other than 2 A socket outlets	Rated current
(b) 2 A socket outlets	At least 0.5 A
(c) Lighting outlet*	Current equivalent to the connected load, with a minimum of 100 W per lampholder
(d) Electric clock, electric shaver supply unit, shaver socket outlet (complying with BS 4573), bell transformer, and current-using equipment of a rating not greater than 5 VA	May be neglected
(e) Household cooking appliance	The first 10 A of the rated current plus 30% of the remainder of the rated current plus 5 A if a socket outlet is incorporated in the control unit
(f) All other stationary equipment	British Standard rated current, or normal current

* For discharge lighting a multiplication factor of 1.8 should be applied to allow for control gear currents.

is permitted. For example, a final circuit feeding lighting points connected to a 230 V supply could have 36 tungsten points, each rated at 100 W. This example shows that, although the Regulations would permit such a circuit, as there is no question of overloading or danger, it would prove very inconvenient to the user, and would constitute a very badly planned installation. Even in the smallest domestic installation the lighting points should be distributed on not less than two circuits.

Several of the regulations are fairly liberal in allowing the planner as much freedom as possible, so as not to place unnecessary restrictions when special circumstances have to be met. The planner must have sufficient experience and knowledge of the purpose to which the installation is to be put, to be able to decide what is justified in any particular case.

In actual practice domestic lighting circuits are generally rated at 6 A. This means they are controlled by a 6 A fuse or circuit breaker, wired with 1.0 mm^2 cables (subject to any derating factors), and 0.75 mm^2 flexible cords for pendants. With rewirable fuses it is impossible to

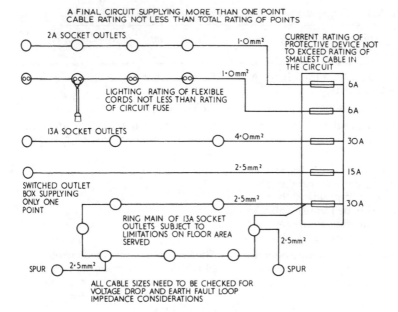

Fig. 5.2   *Final circuits. Diagrammatic representation of the regulations governing the number of points, loading, and sizes of cables and flexibles*

guarantee that the 6 A fuse will not be replaced by a larger fuse wire. If this happens all the precautions which have been taken to ensure correct overcurrent protection will be nullified.

Since the assumed demand of each 2 A socket outlet has to be 0.5 A minimum, it is recommended that 2 A socket outlets to BS 546 should be connected to a circuit not exceeding 6 A and up to ten 2 A socket outlets may be connected to such circuit.

Why the size of flexible cords must be related to the circuit fuse, or overcurrent device   Flexible cords attached to luminaires or to plugs must be of sufficient size so as to be equal to the rating of the fuse which protects the circuit (IEE Regulation 433-02).

When a lighting circuit is protected by a 16 A HBC fuse or circuit breaker, it is important to ensure a sufficiently large flexible cord is used at each pendant. If 0.50 mm^2 flexible cords were fitted to the pendants connected to this circuit, the flexible cord which is only designed to carry 3 A could become dangerously hot in the event of an overload before the circuit protection operates. It is therefore necessary for these flexible cords to be suitable for a continuous current of 16 A (which is 1.5 mm^2).

In industrial installations it is very often necessary to rate final lighting circuits at 16 A. The reason for this is that 300 W, 500 W or even 1000 W lamps may be used and several wired on one circuit. When planning commercial and industrial lighting installations it is not always possible to take advantage of any diversity factors for the distribution circuit as the whole of these lights may be switched on for long periods. This further illustrates the need for knowledge and experience at the planning stage.

To calculate the size of cables or circuits which feed more than one point it is necessary to take the total connected load, refer to the relevant table in Appendix 4 of the IEE Regulations, according to the system of wiring used, then make the necessary adjustment for ambient temperature, grouping and voltage drop.

The rating of the protective device controlling these circuits must not exceed the continuous current-carrying capacity of the circuit conductors (see exemptions for motor circuits and fixed resistive loads).

The current-carrying capacity of the conductors is not necessarily the nominal current stated in the tables. If a larger cable has to be installed due to ambient temperature, grouping or voltage drop, the rating of the cable will be the current the circuit is designed to carry. For example, if a circuit designed to carry 15 A has to be wired with $2.5\,\text{mm}^2$ cable due to derating factors, the rating of this cable will be $I_{tab} \times G$ and not the nominal 24 A as stated in the tables. IEE Regulation 433-02 states the requirements for these design considerations, and examples of this are contained in Chapter 2.

Luminaire track systems to BS EN 60589 are considered to be one point, provided individual luminaires have protecting fuses. To calculate the correct size of cable to use, it is necessary to take the rating of the protective device as the connected load.

Final circuit feeding 13 A sockets to BS 1363

The need for a standard domestic socket outlet with fused plug was first suggested in 1944 by a committee convened by the Institution of Electrical Engineers to study and report on electrical installations in post war buildings (Post War Building Study No. 11).

The multiplicity of types and sizes of socket outlets and plugs in domestic premises had always been a source of annoyance to householders. The committee suggested as a solution to this problem that all existing sockets and plugs should be replaced as opportunity occurred with a standard type of socket outlet.

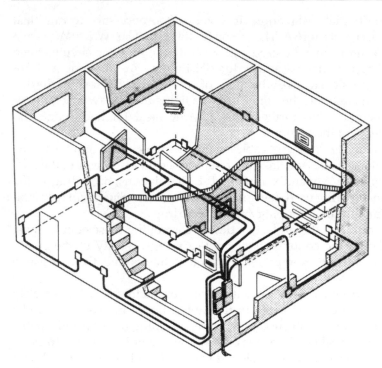

Fig. 5.3   *A typical schematic layout of 13 A socket outlet ring circuits for a small house with a floor area of under 100 m²*

The main advantages of the 13 A socket with fused plug are that any appliance with a loading not exceeding 3 kW (13 A at 230 V) may be connected with perfect safety to any 13 A socket, and under certain conditions an unlimited number of sockets may be connected to any one circuit.

One point which must be borne in mind by the designer is the question of the use of outdoor equipment. IEE Regulation 471-16-01 states that where a socket outlet may be *expected* to supply portable equipment for use outdoors, it shall be protected by a Residual Current Device with a rated residual current not exceeding 30 mA. There are a number of means of achieving this, one of which is illustrated in Fig. 5.4.

Circuit arrangements   Recommendations exist in the IEE *On-site guide* for standard circuit arrangements with 13 A sockets. These permit 13 A sockets to be wired on final circuits as follows (subject to

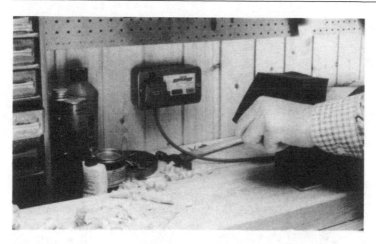

Fig. 5.4 *13 A socket outlets can be obtained which incorporate a 30 mA residual current device. These can be used to satisfy the requirements of the IEE Regulations where RCD protected sockets are required, and are particularly useful in situations where only one or two protected sockets are needed (Ashley and Rock Ltd)*

any derating factors for ambient temperature, grouping or voltage drop):

(a) An unlimited number of socket outlets connected to a final circuit serving a floor area not exceeding $100\,m^2$ wired with $2.5\,mm^2$ PVC insulated cables in the form of a ring (or $1.5\,mm^2$ MI cables in the form of a ring), and protected by a 30 A or 32 A overcurrent protective device.

(b) An unlimited number of socket outlets connected to a final circuit serving a floor area not exceeding $50\,m^2$ with $4\,mm^2$ PVC cables on a radial circuit (or $2.5\,mm^2$ MI cables on a radial circuit) and protected by an overcurrent device of 30 A or 32 A rating.

(c) An unlimited number of socket outlets connected to a final circuit serving a floor area not exceeding $20\,m^2$ with $2.5\,mm^2$ PVC cables on a radial circuit (or $1.5\,mm^2$ MI cables on a radial circuit) and protected by an overcurrent device not exceeding 20 A.

Spurs may be connected to these circuits.

Note that if these standard circuits are used the designer is still responsible for ensuring the circuit is suitable for the expected load. Also the voltage drop, and earth fault loop impedance values are

Fig. 5.5   *Cooker control units, with and without indicator lights (Contactum Ltd)*

suitable and the breaking capacity of the overload protection is sufficiently high.

If the estimated load for any given floor area exceeds that of the protective device given above then the number of circuits feeding this area must be increased accordingly.

For example, a kitchen may have a floor area not exceeding $20\,\text{m}^2$ but the estimated load (apart from the cooker) may well exceed 20 A. In this case the 13 A sockets must be wired on sufficient circuits to carry the known or estimated load. Even in the smallest domestic dwelling it is recommended that at least two circuits should be provided to feed socket outlets.

Spurs

Non-fused spurs   A spur is a branch cable connected to a 13 A circuit. The total number of non-fused spurs which may be connected to a 13 A circuit must not exceed the total number of sockets connected direct to the circuit. Not more than one single or one twin socket outlet or one fixed appliance may be connected to any one spur. Non-fused spurs may be looped from the terminals of the nearest socket, or by means of a joint box in the circuit. The size of the cable feeding non-fused spurs must be the same size as the circuit cable.

Fused spurs   The cable forming a fused spur must be connected to the ring circuit by means of a 'fused connection unit' or spurbox.

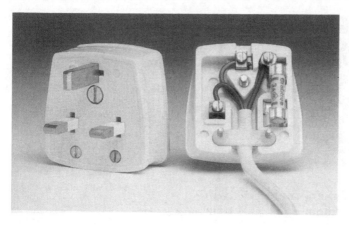

Fig. 5.6   *The 13 A plug, attached permanently to the appliance for which it will be used, can carry a fuse of suitable rating up to 13 A to provide individual protection for that appliance (Crabtree Electrical Industries Ltd)*

The rating of the fuse in this unit shall not exceed the rating of the cable forming the spur, and must not exceed 13 A.

There is no limit to the number of fused spurs that may be connected to a ring. The minimum size of cables forming a fused spur shall be 1.5 mm^2 PVC with copper conductors, or 1.0 mm^2 MI cables with copper conductors.

Fixed appliances permanently connected to 13 A circuits (not connected through a plug and socket), must be protected by a fuse not exceeding 13 A and a d.p. (double pole) switch or a fused connection unit which must be separate from the appliance and in an accessible position.

When planning circuits for 13 A sockets it must always be remembered that these are mainly intended for general purpose use and that comprehensive heating installations, including floor warming, should be circuited according to the connected load, and should not use 13 A sockets.

The Electrical Installation Industry Liaison Committee report recommends a minimum number of 13 A socket outlets that should be installed in domestic premises, as follows:

kitchen	4 twin
living room	6 twin
dining room	3 twin
double bedroom	4 twin
single bed/sitter	4 twin (*Continued overpage*)

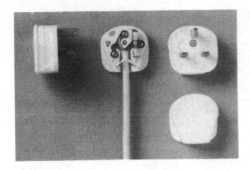

Fig. 5.7   *13 A plug showing the interior with fuse and cable grip arrangement. Under BS 1363 amendment 2, 13 A plugs are required to have sleeved pins (MK Ltd)*

single bedroom/study	3 twin
teenager's room	1 twin
landing/stairs	1 twin
garage	2 twin
hall	1 twin
store/workroom	1 twin
central heating boiler	1 twin.

Fuselinks for 13 A plugs   Special fuselinks have been designed for 13 A plugs, these are to BS 1362 and are standardised at 3 A and 13 A, although other ratings are also available.

Flexible cords for fused plugs

for 3 A fuse $0.50\,mm^2$
for 13 A fuse $1.25\,mm^2$.

All flexible cords attached to portable apparatus must be of the circular sheathed type, and not twin twisted or parallel type. With fused plugs, when a fault occurs resulting in a short circuit, or an overload, the local fuse in the plug will operate, and other socket outlets connected to the circuit will not be affected. It will be necessary to replace only the fuse in the plug after the fault has been traced and rectified.

13 A circuit for non-domestic premises   For industrial, commercial and similar premises the same rules apply as for domestic premises in as much as the final circuit cables must be protected by suitable overcurrent devices.

It is often necessary however, to connect a very large number of sockets to a single circuit, many more than would be recommended for domestic premises. For example in a laboratory it may be necessary to fit these sockets on benches at frequent intervals for the sake of convenience. The total current required at any one time may be comparatively small and therefore a 20 A radial or ring circuit, protected by a 20 A fuse or circuit breaker, and wired with 2.5 mm² PVC cables, could serve a large number of sockets. In this case the area being served must be in accordance with the standard circuit arrangements given in the IEE On-site guide.

Final circuit for socket outlets to BS 196

These are industrial type sockets and plugs, made in three sizes, 5 A, 15 A and 30 A with a maximum voltage of 250 V. As these sockets may be used for any voltage up to 250 V they must be marked with the voltage and current. By a system of keyways the plugs of a given voltage cannot be inserted into a socket of a different voltage. Quite a large selection of supply voltages from 6 V to 250 V are catered for by this system of keyways. All sockets are 2-pole with a scraping earth, and the plugs are non-reversible.

Plugs can be fitted with single-pole or double-pole fuses made to BS 1361 for fuses in the body of the plug, or to BS 196 for plugs with fused pins.

Only single-pole fuses in the live conductor may be used when connected to a normal supply from the mains which has an earthed neutral. Double-pole fusing should be used only when supplied from a double wound transformer (with mid-point of secondary earthed), generator or battery where neither pole is connected to earth.

Standard fuselinks are available as follows:

for 5 A plugs	2 A and 5 A
for 15 A plugs	2 A, 5 A, 10 A and 15 A
for 30 A plugs	10 A, 15 A, 20 A and 30 A.

Plugs with double-pole fuses shall not be interchangeable with those with single-pole or no fuses.

Sockets to BS 196 may be wired on ring or radial circuits, and an unlimited number of sockets may be wired on any one circuit, providing that the overcurrent device does not exceed 30 A and that the size of the conductor is of a current rating not less than that of the overcurrent device (for radial circuits), or for ring circuits not less than 0.67 times the rating of the overcurrent device.

When these sockets are wired on ring circuits, a spur (or branch circuit) may be connected to the ring by the use of a fused connection box fitted with a fuselink not exceeding 15 A with cables capable of carrying this current. Non-fused spurs are not permitted.

Fig. 5.8   *An 8-way 3-phase distribution board fitted with HBC cartridge fuses to BS 88. Common busbars at the base of the panel are for connection of outgoing neutral and circuit protective conductors (Pandelco Ltd)*

The number of sockets connected to each circuit will depend upon the estimated maximum demand. For example, several 30 A sockets may be installed in a workshop to feed a portable welder. These may be connected to a 30 A circuit, when it is known that only one welder will be used at any one time. Such an arrangement would be quite satisfactory. In the same workshop 5 A sockets may be required to feed a large number of portable drills, each with a load of 200 W. As many as 20 of these sockets could be connected to a 20 A circuit. In each case the rating of the overcurrent device must be suitable to protect the conductors installed.

Final circuit for socket outlets to BS EN 60309

These socket outlets are of the heavy industrial type, and are suitable for single phase or three phase with a scraping earth. Fuses are not fitted in the sockets or the plugs. Current ratings range from 16 A to 125 A.

The 16 A sockets whether single- or three-phase, may be wired only on radial circuits. The number of sockets connected to a circuit is unlimited, but the protective overcurrent device must not exceed 20 A. It is obvious that if these 16 A sockets are likely to be fully loaded then only one should be connected to any one circuit. The

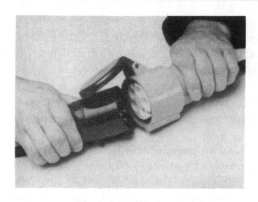

Fig. 5.9   *Industrial plug and socket BS EN 60309-2 (Mk Ltd)*

higher ratings will of course each be wired on a separate circuit. If, on the other hand, a number of these sockets are installed for convenience as given in previous examples, then full advantage should be taken of the flexibility which is permitted.

Due to their robust nature these sockets are often used in industrial installations to feed small 3-phase motors, and if the total estimated load of the motors does not exceed 20 A there is no reason why a considerable number should not be connected to one such circuit.

The same rule which applies to all final circuits must be complied with, which is that the conductors and protective devices must be suitably rated as already explained.

Final circuits feeding fluorescent and other types of discharge lighting

Electric discharge lighting may be divided into two groups: those which operate in the 200 V/250 V range, and the high-voltage type which may use voltages up to 5000 V to earth. The first group includes tubular fluorescent lamps which are available in ratings from 8 W to 125 W, sodium lamps which are rated from 35 W to 400 W, also high pressure mercury vapour lamps rated from 80 W to 1000 W, and other forms of discharge lighting.

The second group includes neon signs and similar means of high voltage lighting.

Low voltage discharge lighting circuits   Regulations governing the design of final circuits for this group are the same as those which apply to final circuits feeding tungsten lighting points, but there are additional factors to be taken into account. The current rating is based upon the 'total steady current' which includes the lamp, and any associated control gear, such as chokes or transformers, and also their harmonic currents. In the absence of manufacturers' data, this can be arrived at by multiplying the rated lamp power in watts by 1.8, and is based on the assumption that the power factor is not less than 0.85 lagging.

For a circuit consisting of twelve 85 W fluorescent lamps, the loading will be $85 \times 12 \times 1.8 = 1836$. For a 230 V supply the current will be $1836\,W/230\,V = 8\,A$.

Some fluorescent lamp circuits (especially the 125 W switch-start type) have very poor characteristics, and manufacturers should be consulted as to the line current which may be expected. In some instances it has been found necessary to multiply the rated power (watts) of the lamp by two. This means that the line current would be $125 \times 2/230\,V = 1.08\,A$ per lamp.

The control gear for tubular fluorescent lamps is usually enclosed in the casing of the luminaire, but for other types of discharge lighting, such as h.p. mercury and sodium, the control gear is sometimes mounted remote from the luminaire. Here it is necessary to check the current which will flow between the control gear and the lamp. The remote control gear must be mounted in a metal box, must be provided with adequate means for the dissipation of heat, and spaced from any combustible materials.

Another disadvantage of locating control gear remote from discharge lamps is that, if a fault develops in the wiring between the inductor and the lamp, the presence of the inductor will limit the fault current so that it may not rise sufficiently to operate the fuse. Such a fault could very well remain undetected. If any faults develop in these circuits this possibility should be investigated.

Circuit switches   Circuit switches controlling fluorescent circuits should be designed for this purpose otherwise they should be rated at twice that of the design current in the circuit; quick-make and slow-break switches must be used. Quick-break switches must not be used as they might break the circuit at the peak of its frequency wave, and cause a very high induced voltage which might flash over to earth.

Polyphase circuits for discharge lighting   In industrial and commercial installations it is sometimes an advantage to split the lighting points between the phases of the supply, and to wire alternate lighting fittings on a different phase so as to reduce stroboscopic effect. When wiring such circuits it is important to provide a separate neutral conductor for each phase, and not wire these on 3-phase

4-wire circuits. The reason for this is that for this type of lighting very heavy currents may flow in the neutral conductors.

Luminaires connected on different phases must be provided with a warning notice DANGER 400 V on each luminaire.

Stroboscopic effect   One of the disadvantages of discharge lighting is the stroboscopic effect of the lamps, although this is not so marked with tubular fluorescent lamps as with some other types. The stroboscopic effect is caused by the fact that the discharge arc is actually extinguished when the voltage reaches zero. This happens twice every cycle, which for a 50 Hz supply is 100 times per second. Although this effect is normally hardly noticeable to the eye, it does sometimes have the effect of making moving objects appear to be standing still, or moving slowly backwards or forwards when viewed under this type of lighting.

This effect can be minimised by connecting alternate lamps or rows of lamps on different phases, so that when one lamp is at zero its neighbour will be somewhere near normal voltage. Another remedy is to use 'lead-lag' two lamp luminaires, or to provide a mixture of fluorescent and tungsten types in close proximity. Some modern fluorescent lamps however operate at high frequency and use phosphors which reduce the stroboscopic effect due to their light retention properties.

High-voltage discharge lighting circuits   High voltage is defined as a voltage in excess of low voltage i.e. over 1000 V a.c. The IEE Regulations generally cover voltage ranges only up to 1000 V a.c., but Regulation 110-01 also includes voltages exceeding low voltage for equipment such as discharge lighting and electrode boilers.

Discharge lighting at high voltage consists mainly of neon signs, and there are special regulations for such circuits. The installation of this type of equipment is usually carried out by specialists.

The equipment must be installed in accordance with the requirements of British Standard BS 559, 'Specification for electric signs and high-voltage luminous-discharge-tube installation'. A linked Switch must be provided to disconnect all live conductors to a step-up transformer. A typical arrangement for a high-voltage discharge lighting circuit is shown in Fig. 5.10.

Final circuits feeding motors

Final circuits feeding motors need special consideration, although in many respects they are governed by the regulations which apply to

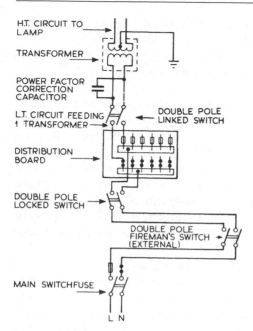

H.T. CIRCUIT TO LAMP

TRANSFORMER

POWER FACTOR CORRECTION CAPACITOR

L.T. CIRCUIT FEEDING 1 TRANSFORMER

DOUBLE POLE LINKED SWITCH

DISTRIBUTION BOARD

DOUBLE POLE LOCKED SWITCH

DOUBLE POLE FIREMAN'S SWITCH (EXTERNAL)

MAIN SWITCHFUSE

L N

Fig. 5.10   *Typical circuit feeding h.t. electric discharge lamps*

other types of final circuits. The current ratings of cables in a circuit feeding a motor must be based upon the full load current of the motor, although the effect of starting current will need to be considered if frequent starting is anticipated (IEE Regulation 552-01-01). Every electric motor exceeding 0.37 kW shall be provided with control equipment incorporating protection against overload of the motor. Several motors not exceeding 0.37 kW each can be supplied by one circuit, providing protection is provided at each motor.

Need for isolators   Every motor shall be provided by means, suitably placed, so that all voltage may be cut off from the motor, starter and all apparatus connected thereto (IEE Regulation 476-02-03). This means that an isolating switch must be provided to cut off pressure from the motor starter, the motor, and all other apparatus, control circuits and instruments used with them.

Alternatively the means of isolation could be fuses (Table 5.3) in a distribution board providing the distribution board is 'suitably placed'; which means in the vicinity of the motor starter, and certainly in the same room. Where several motors are provided for one machine (such as a crane) then only one isolator need be provided to control the group of motors. However, a means of switching off for mechanical maintenance has to be provided (IEE Regulation 462-01-01).

Table 5.3   *HBC fuses for motor circuits*

Maximum full load current (A)	A.C. motors direct-on-line starting (A)	D.C. motors and a.c. star-delta or slipring starting (A)
1.5	5	5
3	10	5
5	15	10
7.5	30	15
10	30	30
15	60	30
20	60	60
30	100	60
50	150	100
70	200	150
100	300	200
150	400	300
170	500	400
200	600	400
250	–	500
300	–	600

The table gives the recommended sizes of HBC fuses for average conditions.

Note that for circuits other than those controlling motors and ring circuits the rating of the fuse shall not be larger than the rating of the smallest cable in the circuit.

Fusing factor
The ratio between the rating and the fusing current of an HBC fuse will depend upon its design.

Fuse manufacturers will give advice for meeting special circumstances.

All isolators must be 'suitably placed' which means they must be near the starter, but if the motor is remote and out of sight of the starter then an additional isolator must be provided near the motor. All isolators, of whatever kind, should be labelled to indicate which motor they control.

The cutting off of voltage does not include the neutral in systems where the neutral is connected to earth. For the purposes of mechanical maintenance, isolators enable the person carrying out maintenance to ensure that all voltage is cut off from the machine and the control gear being worked upon, and to be certain that it is not possible for someone else to switch it on again inadvertently. Where isolators are located remote from the machine, they should have removable or lockable handles to prevent this occurrence.

Motor starters   It is necessary that each motor be provided with a means of starting and stopping, and so placed as to be easily worked by the person in charge of the motor. The starter controlling every motor must incorporate means of ensuring that in the event of a drop in voltage or failure of the supply, the motor does not start automatically on the restoration of the supply, where unexpected re-starting could cause danger. Starters usually are fitted with undervoltage trips, which have to be manually reset after having tripped.

Every motor having a rating exceeding 0.37 kW must also be controlled by a starter which incorporates an overcurrent device with a suitable time lag to look after starting current (IEE Regulation 552-01-02). These starters are generally fitted with thermal overloads which have an inherent time lag, or with the magnetic type which usually have oil dashpot time lags. These time lags can usually be adjusted, and are normally set to operate at 10% above full load current. Electronic protective relays are also available now and these provide a fine degree of protection.

Rating of protective device   IEE Regulation 473-01-02 states that the overcurrent protective device may be placed along the run of the conductors (providing no branch circuits are installed), therefore the overcurrent protective device could be the one incorporated in the starter, and need not be duplicated at the commencement of the circuit.

Short-circuit protection must be provided to protect the circuit, and shall be placed where a reduction occurs in the value of the current-carrying capacity of the conductors of the installation (i.e. such as in a distribution board). The device may, however, be placed on the load side of a circuit providing the conductors between the point where the value of the current-carrying capacity is reduced and the position of the protective device do not exceed 3 m in length and providing the risk of fault current, fire and danger to persons is reduced to a minimum (IEE Regulation 473-02-02) (Fig. 5.14).

When motors take very heavy and prolonged starting currents it may well be that fuses will not be sufficient to handle the starting current of the motor, and it may be necessary to install an overcurrent device with the necessary time delay characteristics, or to install larger cables.

With 3-phase motors, if the fuses protecting the circuit are not large enough to carry the starting current for a sufficient time, it is possible that one may operate, thus causing the motor to run on two phases. This could cause serious damage to the motor, although most motor starters have inherent safeguards against this occurrence.

The ideal arrangement is to back up the overcurrent device in the motor starter with HBC fuselinks which have discriminating characteristics which will carry heavy starting currents for longer periods than the overload device. If there is a short circuit the HBC fuses will

Fig. 5.11   *Suspended ceilings are regularly encountered in commercial premises. The matrix is designed to accept air conditioning, fire detection and lighting fittings, as seen in this view. The luminaires are Dextra compact fluorescent fittings with reflectors which give indirect as well as direct illumination (W.T. Parker Ltd)*

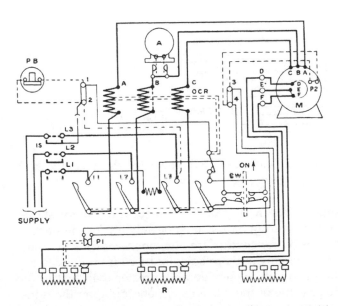

Fig. 5.12   *Stator-rotor starter for 3-phase induction motor. Note that the current in the cables connecting the rotor to starting resistances may be greater than the stator current. These cables must be rated accordingly (Allen West & Co. Ltd)*

Fig. 5.13   *A variety of final circuits in an industrial situation. Socket outlets to BS 1363, BS 196 and BS EN 60309 are all to be seen, the latter being protected by a residual current device. Note also the panel (top right) containing eight contactors controlling lighting circuits. The contactors are switched by eight of the lighting switches below (BP Ltd)*

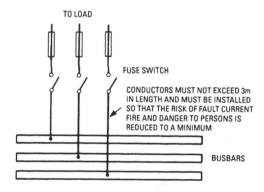

TO LOAD

FUSE SWITCH

CONDUCTORS MUST NOT EXCEED 3m IN LENGTH AND MUST BE INSTALLED SO THAT THE RISK OF FAULT CURRENT FIRE AND DANGER TO PERSONS IS REDUCED TO A MINIMUM

BUSBARS

Fig. 5.14   *Illustrating IEE Regulation 473-02-02*

operate and clear the short circuit before the short circuit kVA reaches dangerous proportions.

Slip-ring motors   The wiring between a slip-ring motor starter and the rotor of the slip-ring motor must be suitable for the starting and load conditions. Rotor circuits are not connected directly to the supply, the current flowing in them being induced from the stator.

The rotor current could be considerably greater than that in the stator; the relative value of the currents depending upon the transformation ratio of the two sets of windings.

The cables in the rotor circuit must be suitable not only for full load currents but also for starting currents. The reason is that, although heavy starting currents may only be of short duration (which the cables would easily be able to carry), if the cables are not of sufficient size to avoid a voltage drop this could adversely affect the starting torque of the motor.

The resistance of a rotor winding may be very low indeed, and the resistance in the rotor starter is carefully graded so as to obtain maximum starting torque consistent with a reasonable starting current.

If cables connected between the rotor starter and the rotor are fairly long and restricted in size, the additional resistance of these cables might even prevent the motor from starting. When slip ring motors are not fitted with a slip ring short-circuiting device, undersized rotor cables could cause the motor to run below its normal speed.

Before wiring rotor circuits always check the actual rotor currents, and see that the cables are of sufficient size so as not to adversely affect the performance of the motor.

Emergency switching   IEE Regulation 537-04-01 states that 'means of interrupting the supply for the purpose of emergency switching shall be capable of cutting off the full load current of the relevant part of the installation'.

In every place in which machines are being driven by any electric motor, there shall be means at hand for either switching off or stopping the machine if necessary to prevent danger. The 'means at hand' could be the isolating switch or the starter, if it is placed sufficiently near the operator to enable the machine to be stopped quickly, but if these are located a distance from the operator then either a STOP button or some mechanical means must be provided to enable the operator to stop the machine.

Generally it is desirable to stop the motor which drives the machine, and if the 'means at hand' is not near the operator then STOP buttons should be provided at suitable positions (Fig. 5.15), and one must be located near the operator, or in the case of large machines, the operators.

Stop buttons should be of the lock-off type so that the motor cannot be restarted by somebody else until such time as the stop button which has been operated is deliberately reset.

In factory installations it is usual to provide stop buttons at vantage points throughout the building to enable groups of motors to be stopped in case of emergency. These buttons are generally connected so as to control a contactor which controls a distribution board, or motor control panels.

For a.c. supplies stop buttons are arranged to open the coil circuit of a contactor or starter. For d.c. supplies the stop buttons are wired to short circuit the hold-on coil of the d.c. starter.

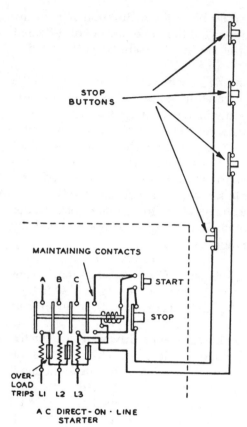

STOP
BUTTONS

MAINTAINING CONTACTS

START

STOP

A   B   C

OVER-
LOAD
TRIPS LI L2 L3

A C DIRECT - ON · LINE
STARTER

Fig. 5.15   *Safety precaution. Means must be at hand for stopping machines driven by an electric motor. One method of doing this is to fit remote STOP buttons at convenient positions*

Reversing three-phase motors   When three-phase motors are connected up for the first time it is not always possible to know in which direction they will run. They must be tested for direction of rotation. If the motor is connected to a machine do not start it if there is a possibility that the machine may be damaged if run in the wrong direction.

If the motors run in the wrong direction it is necessary only to change over any two wires which feed the starter (L1, L2, L3).

In the case of a star delta starter on no account change over any wires which connect between the starter and the motor because it is possible to change over the wrong wires and cause one phase to oppose the others.

For slip ring motors it is necessary only to change over any two lines feeding the starter, it is not necessary to alter the cables connected to the rotor.

To reverse the direction of single-phase motors it is generally necessary to change over the connections of the starting winding in the terminal box of the motor.

Lifts   Electrical installations in connection with lift motors must comply with BS EN 81-2.

The actual wiring between the lift control gear and lift is carried out by specialists, but the designer of the submain needs to comply with the requirements of BS EN 81-2. The power supply to a lift or to a lift room, which may control a bank of lifts, must be fed by a separate distribution cable from the main switchboard. The cable must be of such a size that for a three-phase 400 V supply the voltage drop must not exceed 10 V when carrying the *starting* current of the lift motor or the motor generator. This is the usual maximum volt drop specified by lift manufacturers.

The main switchgear should be labelled LIFTS and in the lift room circuit breakers or a distribution board must be provided as required by the lift manufacturers.

The supply for the lift cage light must be on a separate circuit. It is usual to provide a local distribution board in the lift motor room and the lights controlled by a switch in the lift motor room. These cables must be entirely separate from the cables feeding the power supply to the lift. These lights should be connected to a maintained/emergency supply, so that in the event of mains failure the lights in the lift cage are not affected. Alarm systems should also be connected to a maintained/emergency supply or from a battery.

Cables other than those connected to lift circuits must not be installed in lift shafts, but cables connected to lift circuits need not necessarily be installed in lift shafts.

In large buildings a lift designated as a fireman's lift could be on a separate circuit, so that in the event of a fire the supply to this lift is maintained when other supplies are switched off. Otherwise it could be fed from the main lift riser, in which case the riser should be controlled by a fuseswitch or circuit breaker on the main switchboard which bypasses the main isolator for the building.

Final circuits feeding cookers

In considering the design of final circuits feeding a cooker, diversity may be allowed. In the household or domestic situation, the full load current is unlikely to be demanded and guidance is given in Table 5.2.

If a household cooker has a total loading of 8 kW the total current at 230 V will be 34.8 A, but when applying the diversity factors in Table 5.2 the rating of this circuit will be:

first 10 A of the total rated current	=	10.0 A
30% of the remainder	=	7.4 A
5 A for socket	=	5.0 A
Total	=	22.4 A

Therefore the circuit cables need only be rated for 22.4 A and the overcurrent device of similar rating.

Cookers must be controlled by a switch which must be independent of the cooker. In domestic installations this should preferably be a cooker control unit (Fig. 5.5) which must be located within 2 m of the cooker and at the side so that the control switch can be more easily and safely operated. Cooker control units often incorporate a 13 A socket outlet with a fused plug, although a cooker switch alone is sometimes installed now.

Pilot lamps within the cooker control unit need not be separately fused. Reliance must not be placed upon pilot lamps as an indication that the equipment is safe to handle.

6 Special types of installation

Certain types of installation demand special consideration when designing and installing the electrical equipment. Part 6 of the 16th edition of the IEE Wiring Regulations sets out the specific needs of some types of special installation and the IEE Regulations contained therein supplement or modify the other parts of the IEE Regulations. Details of some of the installations and locations referred to are covered in this chapter.

Part 6 of the IEE Regulations also covers requirements for:

Restrictive conductive locations
Earthing requirements for the installation of equipment having high protective conductor currents.

This book does not go into details of these installations and locations but information may be obtained by reference to IEE Regulations Sections 606 and 607, and the IEE Books of Guidance Notes.

Bath and shower rooms

In rooms containing a bath or shower the risk of electric shock is increased due to the fact that the body is in contact with earth and, as a result of being wet, has reduced electrical body resistance. IEE Regulations Section 601 details the requirements and specifies zones in bath and shower rooms with restrictions on equipment which may be fitted.

Limitations exist as to the forms of shock protection which may be used and supplementary equipotential bonding must be provided. No switches, socket outlets or other electrical equipment may be installed unless certain conditions are met.

Light fittings should ideally be of the totally enclosed type. Alternatively lampholders must be shrouded or fitted with a protective shield to BS EN 61184. Switches must be placed such that they are

Fig. 6.1 *Shaver units which incorporate an isolating transformer to BS EN 61558 are permitted in bathrooms (Crabtree Electrical Industries Ltd)*

Fig. 6.2 *Cable operated light switches for use in bathrooms can be obtained with or without neon indicators (Crabtree Electrical Industries Ltd)*

inaccessible to a person in the bath or shower, unless they are supplied by SELV (extra low voltage), not exceeding 12 V, or are part of a shaver unit incorporating an isolating transformer to BS EN 61558. Similarly with socket outlets, none are permitted unless supplied by SELV, not exceeding 12 V.

Cord operated switches may be used provided the switch itself is inaccessible and, if a shower cubicle is fitted in a room other than a bath or shower room, then any socket must be at least 2.5 m from the cubicle. Other special requirements apply to bath or shower room heaters.

Swimming pools, paddling pools and hot air saunas

As with bath and shower rooms, increased precautions against electric shock are required in these locations and in certain specified zones within or near them. IEE Regulations Sections 602 and 603 set out the details for swimming pools and hot air saunas respectively. Requirements include the provision of barriers with appropriate degrees of protection in accordance with BS EN 60529, placing certain equipment outside specified zones, provision of SELV (extra low voltage) supplies not exceeding 12 V, and constraints on the type of wiring systems which may be used. In the case of hot air saunas, provision to avoid the overheating of electrical equipment must be made.

There is a duty upon the designer to extend the assessment of general characteristics. Specific examination of these areas and the way in which they may be used must be made. Additional information is contained in the IEE Regulations themselves and in the IEE books of Guidance Notes.

Off-peak heating

The supply authorities offer cheaper tariffs for off peak loads. As electricity cannot be stored in quantity, they can improve their load factor by encouraging the use of electricity during these periods. Although electricity itself cannot be stored, the heat generated by electricity can be stored in water, and massive objects such as concrete and similar materials.

In large premises water can be heated during the night at lower tariff charges and used during the daytime. For domestic installations water storage heaters have been specially designed with sufficient capacity to enable normal requirements for hot water to be heated during off-peak hours in the same way.

For space heating of buildings an increasing number of off-peak heating installations are being installed. One common method is the installation of the storage radiator rated at between 1 and 6 kW. The cubic capacity and the weight of these are relatively large to enable them to store sufficient heat during the night to provide the required heat emission during the day. There are other types available which are lagged, and have a controlled output, usually by means of a fan which can be switched on when heat is required. These are called storage fan heaters.

Off-peak heaters have to be wired in circuits back to a separate main switch, which is controlled by a contactor and time switch.

When the heaters are fitted with fans to control the heat output, these fans have to be wired on the ordinary circuits so that they are not switched off during the daytime.

A notice should be fixed on these heaters which gives a warning that although the heater is OFF, the fan may still be alive.

Underfloor and ramp heating

An alternative method of taking advantage of off-peak electricity tariffs is the use of underfloor warming.

This system takes advantage of the thermal storage properties of concrete, and utilises the concrete floors of buildings for storing heat generated by electric currents.

The usual method is to embed heating cables in the floors; these are usually laid on the floor slab, and then covered with a 70 to 75 mm layer of concrete screeding (see Figs. 6.3 and 6.4).

Great care must be taken in laying these cables, and in protecting them against accidental damage during screeding.

Special instruments are available for connecting up to the cables during installation which will give an audible warning as soon as any cable becomes damaged, and thus enables the damage to be rectified before it is too late.

The ends of the heating cables are taken to junction boxes near skirting level, and then connected by ordinary wiring to the distribution board, including connections to room thermostats, etc.

The design of these installations is a specialised job, and many factors have to be taken into account, including heat losses to the room below.

As in the case of block storage heaters, the circuits feeding underfloor warming must be taken back to separate distribution boards and controlled by a contactor and time switch.

When floor warming cables are installed it is necessary to ensure that conduits and other pipes are kept well clear of these cables.

It is often advisable to install MI cables in floors when heating cables are in close proximity; and it is good practice to use MI cables to feed the heating cables.

Fig. 6.3 *Installation of cables for floor heating (ESWA Ltd)*

Fig. 6.4 *Floor warming being installed in the Orangery at St Johns Wood (DE-VI Electroheat Ltd)*

IEE Regulation 554-06 deals with floor warming installations. IEE Table 55C gives maximum operating conductor temperatures for floor warming cables. Conductors insulated with PVC or enamel have a maximum temperature of 70°C, those with heat resisting insulation have a maximum of 85°C, and for MI cables with a copper sheath without PVC covering the maximum is up to 150°C.

When installed in large floor areas, suitable precautions must be taken to provide expansion joints in the cables where expansion is normally to be expected, and where these are provided in the build-ing structure.

Where underfloor warming is installed in special areas such as bath or shower rooms, swimming pools or paddling pools, add-itional supplementary equipotential bonding is required, which is connected to a metallic grid covering the heating units.

Apart from the use of underfloor heating in buildings to gain advantage of off-peak electricity tariffs, another use being made of this type of installation is in ramp heating for motor vehicles. The method of design and installation is the same as described above, but in this case, of course, the supply is derived from the normal feeds, and not via the control of an off-peak time switch. The ramp heating equip-ment is provided to enable vehicles to move freely in conditions of snow and ice, so the supply must be available on demand. It is pos-sible to obtain control equipment which monitors precipitation as well as temperature, and quite sophisticated automatic control of such heating installations can be obtained.

Installations in hazardous areas

Hazardous areas mainly consist of places where potentially flam-mable materials are present. This includes spraying and other paint-ing processes which involve the use of highly flammable liquids, installations associated with petrol service pumps, and inspection pits in garages.

Electricity at Work Regulation 6 states that electrical equipment which may reasonably foreseeably be exposed to hazardous conditions 'shall be of such construction or as necessary protected as to prevent, so far as is reasonably practicable, danger arising from such exposure'.

The Fire Offices Committee have issued recommendations for electrical installations in connection with highly flammable liquids used in paint spraying. The Ministry of Fuel and Power also stated conditions for the granting of petroleum spirit licences in respect to electrical equipment. These conditions require that petrol pumps shall be of flameproof construction, so also shall be switchgear and other electrical control gear. Luminaires within the pump equipment shall be of flameproof construction, but those mounted outside the pump casing shall be of totally enclosed design in which the lamp is protected by a well glass or other glass sealed to the body of the luminaire so as to resist the entry of petroleum spirit vapour. The wiring shall be carried out by insulated cables enclosed in heavy gauge galvanised solid-drawn steel conduit. Conduit boxes within the pump equipment shall be of flameproof construction and galvanised. Alternative wiring may consist of MI cables, copper sheathed with flameproof glands.

The supply circuits for each pump shall be separately protected with overcurrent protection, and these protective devices shall not be situated within, or on, the pump housing.

BS 229 and BS 889 give details of the flameproof equipment which must be used in petrol pump enclosures.

It should be noted that in petrol filling stations the installation of electrical supply to the PME (protective multiple earthing) arrangement is specifically prohibited under Health and Safety Guidance Note HS (G) 41. This is because with significant amounts of underground equipment such as supply pipes and storage tanks it is necessary to prevent fault currents flowing to earth in this hazardous environment.

Where explosive dusts are likely to be present, flameproof equipment and circuit systems must be used, but the luminaires and conduit fittings and other electrical equipment must also be fitted with dust-tight gaskets to prevent the entry of explosive dusts. Without these dust-tight gaskets the ordinary flameproof accessory could breathe in explosive dusts between the machined surfaces when changes in temperatures occur.

Hazardous areas are classified as follows:

Zone 0   in which an explosive gas/air mixture is continuously present for long periods.

Zone 1   in which an explosive gas/air mixture is likely to occur in normal operation.

Zone 2   in which an explosive gas/air mixture is not likely to occur in normal operation, and if it occurs it will exist only for a short time.

Electrical installations for caravan parks and caravans

Definitions   The IEE Regulations define a caravan as 'a trailer leisure accommodation vehicle, used for touring, designed to meet the requirements for the construction and use of road vehicles'. The IEE Regulations also contain definitions of motor caravan and leisure accommodation vehicles.

A caravan park is an area of land that contains two or more caravan pitches. Additional IEE Regulations for caravans, motor caravans and caravan parks are contained in IEE Section 608, division one and division two respectively.

Caravan parks

As electrical installations on caravan sites are extremely vulnerable to possibilities of shock due to their temporary nature, special regulations have been made. A socket outlet controlled by a switch or circuit breaker protected by an overcurrent device and a residual current device shall be installed external to the caravan, and shall be enclosed in a waterproof enclosure, it shall be non-reversible with provision for earthing BS EN 60309-2.

At these positions a warning notice must be provided giving details of voltage, whether a.c. or d.c., the frequency and the maximum permissible load.

External installations on caravan parks, although some may only be of a temporary nature, must be carried out strictly to the IEE Regulations in general, and the Electricity Supply Regulations.

Caravan installations

All mobile caravans shall receive their electrical supply by means of a socket outlet and plug of at least 16 A capacity, with provision for earthing BS EN 60309-2. These sockets and plugs should have the phase, neutral and earth terminals clearly marked, and should be sited on the outside of the caravan. They should be connected to the main switch inside the caravan by cables 25 m in length. A notice must be fixed near the main switch inside the caravan bearing indelible characters, with the text as given in IEE Regulation 608-07-05. This notice gives instructions to the caravan occupier as to precautions which are necessary when connecting and disconnecting the caravan to the supply. It also recommends that the electrical installation in the caravan should be inspected and tested at least once every 3 years.

Other recommendations are that all wiring shall be rubber or PVC sheathed, and shall not come in contact with any metal work. Cables shall be firmly secured by non-corrosive clips at intervals not exceeding 250 mm for horizontal runs and 400 mm for vertical runs. Luminaires shall be mounted direct on the structure, flexible pendants must not be installed. Protective conductors shall be insulated from the metalwork of the structure, and should preferably be incorporated in the cable containing the circuit conductors. Protective conductors shall terminate at an earthing terminal insulated from the structural metalwork, and placed near the main switchgear. This earthing terminal shall be connected to the earthing pin of the caravan inlet.

IEE Regulation 608-03 specifies which methods may be used for protection against indirect contact and characteristics of protective devices and cables.

Installations on construction sites

Temporary electrical installations on building and construction sites are necessary to enable lighting and power to be provided for the various trades engaged on the site. These temporary installations need to be of a very high standard owing to the exceptional hazards which obtain.

Temporary installations are required to comply with the Electricity at Work Regulations 1989 issued by the Health and Safety Commission, also British Standard BS 7375 and BS 4363. IEE Regulations Section 604 also applies.

The Electricity at Work Regulations 1989, which apply to permanent installations, also apply to temporary installations on construction sites, so these temporary installations must be of the same standard as those for other installations. Not so many years ago regulations for electrical installations on building sites were practically non-existent. Portable lamps consisting of brass lampholders with a twisted 2-core flexible cord were in frequent use, and many fatal accidents occurred as a result.

The use of these in the vicinity of earthed metal or damp floors presented a real hazard, even when connected to extra-low voltage supplies. Apart from the danger of shock there is a danger to the eyes should the lamp be accidentally broken. All portable handlamps must be properly insulated and fitted with a guard.

Construction site lighting is necessary to cover the following requirements:

(1) Lighting of working areas, especially internal working areas where there is no natural light, with a minimum intensity of 20 lux.
(2) Walkways, especially where there are uneven floors, minimum intensity 5 lux.
(3) Escape lighting, along escape routes, this lighting to be from a supply separate from the mains supply, usually battery operated, minimum intensity 5 lux.

(4) Emergency lighting. This to be in accordance with BS 5266 Part 1 and to come on automatically in the event of mains failure. Usually battery operated or from a generating set. Minimum intensity 5 lux.

High level fixed lighting could be taken from 230 V mains supply, but low level and portable lighting should be 110 V with centre point earthed via a double wound transformer.

In vulnerable situations, such as in damp areas, tanks, etc., the voltage should be reduced to the SELV voltage (50 V a.c). Residual current circuit breakers should control these circuits, with a monitoring unit (sometimes referred to as an earth proving unit) to ensure that the earth connection is intact.

Power for tower cranes mixers and other motors over 2 kW is usually supplied from a 400 V mains supply. Sockets for portable tools should be to BS 4363 and BS EN 60439-4, fed from a 110 V supply via a double wound transformer. A diagram showing a typical distribution layout is given in Fig. 6.5.

IEE Regulations Section 604 includes a number of additional and amended regulations which apply to construction sites. Because of the increased risk of hazards which exist in these locations some tightening of requirements, particularly with regard to shock protection is called for.

If the TN supply system is in use with eebad (earth equipotential bonding and automatic disconnection of supply) used for protection, the disconnection times required are considerably less than those required in ordinary installations. IEE Regulation 413-02-10, for example, is replaced by IEE Regulation 604-04-03, which requires a 0.2 second disconnection time. Similar levels of change apply in the use of TT and IT systems and the IT system itself is to be avoided as a means of supply if at all possible. If an IT system must be used, then permanent earth fault monitoring must be in place.

A number of BS (British Standards) apply to installations on construction sites. BS 4363, BS 5486 and BS EN 60439-4 cover electricity supplies and equipment, BS EN 60309 applies to plugs, sockets and cable couplers, and certain minimum standards for enclosures are required consistent with BS EN 60529: 1992.

Electrical conductors must not be routed across roadways without adequate mechanical protection, and all electrical circuits must have isolators at each supply and distribution point which are capable of being locked in the off position. Additional useful information is contained in Section 604 of the IEE Regulations.

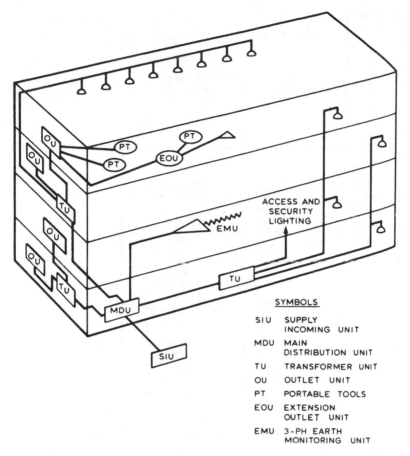

Fig. 6.5   *Typical arrangement of distribution units for construction sites (from BS 7375)*

Emergency supplies to premises

The need for emergency supplies in factories, commercial buildings, hospitals, public buildings, hotels, multi-storey flats, and similar premises is determined by the fire prevention officer of the local authority concerned, and is also related to the need to provide a minimum level of continuity of supply.

The object of emergency lighting is to provide adequate illumination along escape routes within 5 seconds of the failure of normal lighting. BS 5266 deals with the emergency lighting of premises other than cinemas and certain other premises used for entertainment.

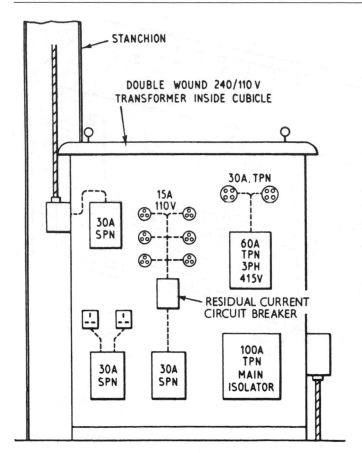

Fig. 6.6   *Load centre for temporary installation*

If the recommendations of BS 5266 Part 1 are complied with it is almost certain that the emergency lighting system will be acceptable to the local 'enforcing authority'.

The Fire Protection Act of 1971 indicates the need for escape lighting, but does not make any specific demands. IEE Regulation 313-02 mentions that any emergency supplies required by the enforcing authority shall have adequate capacity and rating for the operation specified. BS 5266 Part 1 recommends that emergency lighting be provided in the following positions:

(1) Along all escape routes towards and through all final exits, including external lighting outside all exits.
(2) At each intersection of corridors, and at each change of direction.

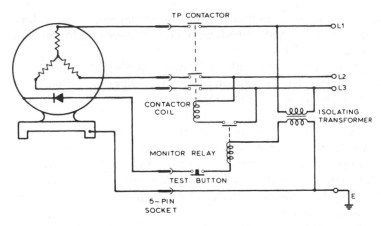

Fig. 6.7 *Typical monitoring circuit for a 3-phase motor fed via a 5-pin socket and flexible cable, for construction sites (from BS 7375)*

(3) On staircases to illuminate each flight of stairs, and near any change of floor level.

(4) To illuminate all exit signs, directional signs, fire alarm contacts, and fire fighting appliances. (*Note:* The illumination of signs may be either from within or external to the sign.)

(5) All lifts in which passengers may travel.

(6) All toilet areas which exceed eight square metres.

(7) Over moving staircases or walkways (i.e. escalators and travelators) as if they were part of the escape route.

(8) Control, plant, switch and lift rooms.

Emergency lighting must come into operation within 5 s of the failure of the normal lighting, and must be capable of being maintained for a period from 1 to 3 hours (according to the requirements of the local 'enforcing authority'). The level of illumination should be not less than 0.2 lux measured at floor level on the centre line of the escape route.

Along corridors it is recommended that the spacing of lighting luminaires should have a maximum ratio of 40:1 (i.e. distance between luminaires and mounting height above floor level) and, of course the illumination levels above must also be met.

Emergency lighting systems

Alternative methods of providing emergency lighting are as follows:

(1) Engine driven generating plants, capable of being brought on load within 5 s.

(2) Battery powered systems, utilising rechargeable secondary batteries, combined with charger, centrally located to serve all emergency lights.
(3) Signs or luminaires with self-contained secondary batteries and charger. The battery after its designed period of discharge being capable of being re-charged within a period of 24 hours.

Circuits feeding luminaires or signs with self-contained batteries shall be continuously energised, and steps must be taken to ensure that the supply is not inadvertently interrupted at any time. Switches or isolators controlling these and other emergency lighting circuits must be placed in positions inaccessible to unauthorised persons, and suitably identified.

Standby supplies

In addition to emergency escape lighting, it is very often desirable to provide 'standby supplies' which will come into operation in the event of a failure of the supply. This lighting is intended to provide sufficient illumination to enable normal work to be carried on. It is very often necessary where there are continuous processes, which must not be interrupted, and in computer installations. In these cases it is also necessary to provide standby power supplies to enable the processes to continue.

In large installations, it is usual to provide standby diesel driven alternators for essential services. It is unlikely these will be able to

Fig. 6.8   *Automatic emergency lighting unit. The unit has a self-contained rechargeable battery, and control gear which detects mains failure and energises the emergency lights (BP Ltd)*

supply the full load of the building, and thus some load shedding will be necessary. Because the occupants of the building will not normally carry out this function, some special circuit provision should be made. The essential loads should ideally be determined at the design stage, and separate distribution arrangements made from the main switchboard. The changeover switching arrangements are usually automatic, and the circuits must be arranged in such a way that the standby power supplies feed only the essential load distribution network.

System of wiring

Wiring to emergency supplies must comply with the IEE Regulations and BS 5266. Recommended systems of wiring are MI cables, PVC/armoured cables, FP cables, PVC or elastomer insulated cables in conduit or trunking. In certain installations the use of plastics

Fig. 6.9   *The main switchboard of an agricultural installation. Individual fuse switches feed the Dutch barn, silos and grain store. Note that the installation is three phase, and is protected by a residual current circuit breaker (William Steward & Co Ltd)*

conduit or trunking is prohibited, and the enforcing authority should be consulted on this.

All wiring for emergency lighting and fire alarms if enclosed in conduit or trunking must be segregated from all other wiring systems (see IEE Regulation 528-01-04). When trunking is used the emergency lighting and fire alarm/circuits, cables must be segregated from all other cables by a continuous partition of non-combustible material.

Multicore cables should not be used to serve both emergency and normal lighting (BS 5266).

Agricultural and horticultural premises

Agricultural installations, which include buildings accessible to livestock, require very special consideration. Horses and cattle have a very low body resistance which makes them susceptible to electric shock at voltages lower than 25 V a.c. It is recommended that electrical equipment installed in these areas should have a degree of protection to at least IP44 and be of Class 2 construction. Switches and other accessories should certainly be placed out of reach of animals, and this generally means that they be placed in enclosures or outside the areas occupied by livestock.

In the case of low-voltage systems it is strongly advised that the circuits should be protected by a residual current circuit breaker, and for socket outlets this requirement is compulsory.

Fig. 6.10   *Fire alarm point and siren wired in MI cable (BP Ltd)*

In vulnerable situations, such as milking parlours, glasshouses and other buildings where water sprays are used or where high humidity exists, it is recommended that extra-low voltage supplies be used, and IEE Regulation 605-02 states that under such circumstances additional precautions be taken in the form of barriers or enclosures or insulation capable of withstanding 500 V a.c. for 60 seconds.

Mains-operated electric fence controllers must comply with BS EN 60335-2-76, and their installation must comply with IEE Regulations 605-14-01 to 605-14-06.

As with other areas of high risk from shock currents, modified arrangements are a requirement of the IEE Regulations regarding the times for automatic disconnection and other associated measures. Supplementary bonding connecting all exposed and extraneous conductive parts must be provided, and this includes any conductive or metal mesh covered floors.

Fire alarms

The design of fire alarm systems does not come within the scope of this book, and it is usual for manufacturers of fire alarm equipment or other specialists in this work to design these installations.

Wiring for fire alarm installations is covered by British Standard BS EN 54-2 and BS EN 54-4, and also the Rules of the Fire Offices Committee for Automatic Fire Alarm Installations. Generally speaking the approved systems of wiring are the same as those for emergency lighting. Wiring installed in conduits or trunking must be segregated from all other types of wiring systems, (except emergency lighting).

Highway power supplies

The 16th edition of the IEE Wiring Regulations includes a separate section, 611, relating to this subject. As with the other sections in Part 6 of the Regulations, the items included supplement or modify the general requirements.

Items under Highway power supplies include altered arrangements for protection against electric shock, and additional requirements for cable identification and protection where conductors are installed in the ground.

Part 2
Practical Work

7 A survey of installation methods

The preceding chapters cover the various regulations governing the control and distribution of supplies, and the design and planning of installation work have been discussed. From now on the practical aspects of electrical installation work will be dealt with.

It is very important that the practical work is carried out correctly.

Cable management systems

Commercial and industrial electrical installations are nowadays generally comprehensive and complex systems, and when installed in new or recently refurbished buildings employ a range of methods in distributing and routeing of electrical circuits. A number of firms are able to supply the full range of equipment needed. This includes a variety of trunking and conduit types, cable tray, cable ladder and such other items as power poles or posts, with which the electrical installer can present a complete and well finished installation.

The collective term used for the variety of methods available is 'cable management systems' and the various elements of them are dealt with separately in this book under the appropriate chapters on conduit, trunking or busbar systems.

Principal types of wiring systems

There are many alternative wiring systems that may be adopted:

(1) PVC single-core insulated cables (70°C) in conduit or trunking
(2) Rubber insulated cables (85°C) in conduit or trunking
(3) PVC insulated and sheathed multicore cables fixed to a surface or concealed
(4) MI or FP cables
(5) PVC single-core sheathed on cleats
(6) PVC multicore sheathed and armoured

Fig. 7.1   *Emergency lighting fittings used in this warehouse stacking system are mounted at floor level on multi-compartment trunking. This enables separation of circuits of different voltage bands (W.T. Parker Ltd)*

    (7) PILC multicore and armoured
    (8) EPR or XLPE multicore armoured
    (9) Busbar systems.

Choice of wiring system

In deciding the type of wiring system for a particular installation, many factors have to be taken into consideration; amongst these are:

    (1) whether the wiring is to be installed during the construction of a building, in a completed building, or as an extension of an existing system
    (2) capital outlay required
    (3) planned duration of installation
    (4) whether damp or other adverse conditions are likely to exist
    (5) type of building
    (6) usage of building
    (7) likelihood of alterations and extensions being frequently required.

Since the relative importance of each of the foregoing factors will vary in each case, the final responsibility must rest with the person planning the installation.

Table 7.1 *Glossary of terms used in cable identification*

Acronym	Meaning	Notes
CPE	Chlorinated polyethylene	
CSP	Chlorosulphonated polyethylene	
CTS	Cab tyre system	See Chapter 12
EPR	Ethylene propylene rubber	
ETFE	Ethylene tetrafluoroethylene	
FP	Fire performance	
FR	Fire retarded	
HOFR	Heat oil resistant and flame retardant	
LDPE	Low density polyethylene	
LSF	Low smoke and fumes	Made to BS 6724
LSOH	Low smoke, zero halogen	No halogens in cable insulation
MI	Mineral insulated	See Chapter 13
NH	Non-halogenated	Made to BS 6724
PILC	Paper insulated lead covered	Often with PVC sheath
PTFE	Polytetrafluoroethylene	
PVC	Polyvinylchloride	Widely used. Suitable for 70°C. Made to BS 6346
RS	Reduced smoke	
SWA	Steel wire armoured	
TCWB	Tinned copper wire braid	
TRS	Tough rubber sheath	
TSWB	Tinned steel wire braid	
VR	Vulcanised rubber	
XLPE	Crosslinked polyethylene	Suitable for 90°C. Made to BS 5467

Frequently a combination of several wiring systems may be used to advantage in any one installation. For example, in an industrial installation the main and submain cables would probably consist of PVC/SWA/PVC cables, or MI copper sheathed cables. The power circuits could use LSF, XLPE or PVC insulated cables in conduit or trunking, or MI cable. The lighting circuits could be carried out with PVC cables in plastic trunking or conduit, or with PVC insulated and sheathed cables fixed to the surface.

Low-voltage wiring
The 16th edition of the IEE Regulations deals with two voltage ranges:

Extra-low voltage	normally not exceeding 50 V a.c. or 120 V d.c. whether between conductors or to earth.
Low voltage	normally exceeding extra-low voltage but not exceeding 1000 V a.c. or 1500 V d.c. between conductors or 600 V a.c. or 900 V d.c. between any conductor and earth.

All conductors shall be so insulated and where necessary further effectively protected be so placed and safeguarded as to prevent danger so far as is reasonably practicable.

Foundations of good installation work
Whatever wiring system is employed there are a number of requirements and regulations which have a general application. These will be mentioned before dealing with the various wiring systems in detail. Regulation 133-01-01 in Chapter 13 of the IEE Regulations says 'Good workmanship and proper materials shall be used'. This is no mere platitude because bad workmanship would result in an unsatisfactory and even a dangerous installation, even if all the other regulations were complied with. Good workmanship is only possible after proper training and practical experience. Knowledge of theory is very necessary and extremely important, but skill can only be acquired by practice. It must always be remembered that the choice of materials, layout of the work, skill and experience all combine to determine the character and efficiency of the installation.

Proper tools   There is an old adage about being able to judge a workman by his tools, this is very true. An electrician must possess a good set of tools if the work is to be carried out efficiently. Moreover, the way in which the electrician looks after his/her tools and the condition in which they are kept is a very sure indication of the class of work likely to be produced. Besides ordinary hand tools, which by custom electricians provide for themselves, there are others such as stocks and dies, electric drills and hammers, bending machines, electric screwing machines which are usually provided by the employer, and which will all contribute to the objective of 'good workmanship'.

Fig. 7.2 *Plastic conduit being embedded in a concrete 'raft'. It is important to ensure that sound joints are made so that the conduit does not part whilst the concrete is being poured (MK Ltd)*

Fig. 7.3 *PVC/SWA/PVC sheathed cables being used as submain cables for distribution of power. Provision has been made for twelve 240 mm^2 4-core cables to be run, and in this view eleven have been fitted (William Steward & Co Ltd)*

Fig. 7.4   *Power posts in conjunction with underfloor trunking systems can provide a very neat office installation (MK Ltd)*

Fig. 7.5   *'Grid' suspension wiring system for supplying electricity by self-supporting cables and connecting boxes, which is suitable for outdoor installations (British Rail)*

Selection of cable runs   Cables and other conductors should be so located that they are not subject to deterioration from mechanical damage, vibration, moisture, corrosive liquids, oil and heat. Where such conditions cannot be avoided, a suitable wiring system must be provided. Four examples are (1) MI cables which will withstand

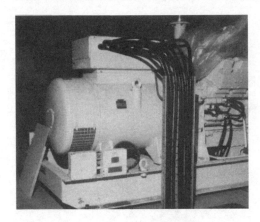

Fig. 7.6 *Cables connecting this standby alternator are mounted on cable tray*

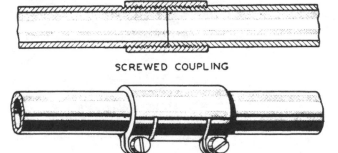

SCREWED COUPLING

GRIP COUPLING

Fig. 7.7 *Couplings for steel conduit. Simple screwed coupling for heavy-gauge screwed conduit and a grip coupling for joining lengths of light-gauge conduit. The latter is used nowadays only for intruder alarm systems*

Fig. 7.8 *Flameproof-weatherproof fittings for hazardous locations. Left, luminaire for 80/125 W. Right, a junction box (Victor Products Ltd)*

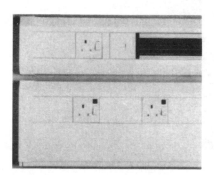

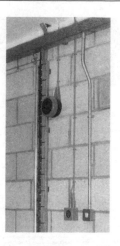

Fig. 7.9   *A busbar distribution system incorporated in trunking and capable of accepting power and data outlets. The data wiring is carried in a separate compartment in the trunking (MK Ltd)*

Fig. 7.10   *The illustration shows a number of wiring systems in use. Cables installed in conduit, trunking and on cable tray are present, and the bond between the cable tray to the trunking can also be seen. The fire alarm and fire alarm bell are wired in FP200 (fire performance) cable (William Steward & Co. Ltd)*

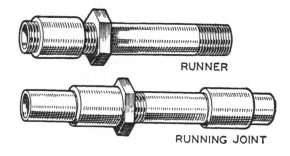

RUNNER

RUNNING JOINT

Fig. 7.11   *Running joint for coupling screwed conduit which cannot be turned.*

water, steam, oil or extreme temperature, (2) LSF or PVC covered MI cables which will withstand, in addition, chemicals corrosive to copper, (3) XLPE up to 90°C and (4) PVC/SWA/PVC sheathed cables will withstand most of these conditions and operating temperatures up to 70°C. Table 7.2 gives the maximum normal operating temperatures of various types of cables.

If operating temperatures exceeding 150°C are encountered then special heat resisting cables must be used, such as varnished glass fibre, for temperatures up to 250°C, and where there are exceptionally high

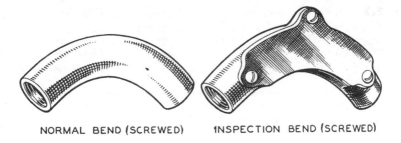

NORMAL BEND (SCREWED) INSPECTION BEND (SCREWED)

Fig. 7.12 *Conduit fittings*

Fig. 7.13 *A cable management system for use in an office installation is shown here. The ceiling void houses non-metallic trunking, and circuits are brought to working height with a power pole. Segregation is provided for power, data and other circuits (MK Ltd)*

temperatures the conductors must be of high melting point materials, such as nickel or chromium copper, or silver plated copper.

It should be noted, as explained in Chapter 2, that the current-rating of cables depends very much upon the ambient temperature in which they are installed. In boiler houses and similar installations where the cables are connected to the thermostats, immersion heaters and other equipment located near or on the boiler, it is usual to carry out most of the wiring with PVC cables in conduit or trunking, or with MI cables, and make the final connection near the boiler by means of a short length of butyl or silicone cable in flexible conduit, these cables being joined with a fixed connector block in a conduit box fitted a short distance away from the high temperature area. The flexible metallic

Table 7.2   *Maximum normal operating temperatures specified for the determination of current-carrying capacities of conductors and cables*

Type of insulation or sheath	Maximum normal operating temperature °C
60°C rubber compounds	60
PVC compound (type TI1, TI2, TM1, TM2, 2 and 6 to BS 6746)	70
MI cables exposed to touch or plastic covered	70
Impregnated paper†	80
85°C rubber compounds	85
PVC compound (types 4 and 5 to BS 6746)	85
HOFR and heavy duty HOFR compounds	85
XLPE	90
Bare MI cables not exposed to touch and not in contact with combustible material	105
150°C rubber compound	150
Varnished glass fibre	180

† Applicable only to cables of voltage rating 600/1000 V.
Note: Where the insulation and sheath are of different materials, the appropriate limits of temperature for both materials must be observed. For cables sheathed with materials not mentioned in the table (e.g. metal), it may be assumed that the limit of temperature specified for the insulation will also ensure a satisfactory sheath temperature.

conduit permits the removal of the thermostat or other device without the need to disconnect the cables (Fig. 7.14).

Cables exposed to corrosive liquids   Where cables are installed in positions which are exposed to acids or alkalis, it is usual to install PVC insulated cables. The use of metal covering should be avoided. Similarly, in the vicinity of sea water, steel conduits or other systems employing ferrous metals should not be used.

Cables exposed to explosive atmospheres   Where conductors are exposed to flammable surroundings, or explosive atmospheres, special precautions have to be taken. The Electricity at Work Regulations 1989 state that electrical equipment which may reasonably foreseeably be exposed to hazardous environments 'shall be of such construction or as necessary protected as to prevent, so far as is reasonably practicable, danger arising from such exposure'. Such installations should comply with BS EN 60079 and BS EN 50014. The Petroleum Regulations Act deals with installations for petrol service pumps and storage depots.

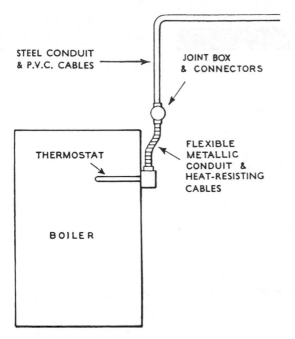

STEEL CONDUIT
& P.V.C. CABLES

JOINT BOX
& CONNECTORS

THERMOSTAT

FLEXIBLE
METALLIC
CONDUIT &
HEAT-RESISTING
CABLES

BOILER

Fig. 7.14   *Wiring cable-feed to boiler thermostat. A short length of heat-resisting cable is used in the vicinity of the boiler where the temperature is high. This is joined to the main PVC cable by a connector*

Preventing the spread of fire   When installing conduits, trunking or cables in any building a very necessary precaution is to avoid leaving holes or gaps in floors or walls which might assist the spread of fire. Vertical cable shafts or ducts could enable a fire to spread rapidly through the building. Any holes or slots which have to be cut in floors or walls to enable cables to pass through must be made good with incombustible material.

Vertical cable ducts or trunking must be internally fitted with non-ignitable fire barriers at each floor level. The slots or holes through which the conduits or trunking pass must be made good at each floor level. The internal non-ignitable barriers not only restrict the spread of fire, but also counteract the tendency for hot air to rise and collect at the top of a vertical duct (Fig. 7.15).

When fitting fire barriers, it is important to select the correct fire stop material. The use of incorrect material may achieve the desired result in preventing spread of fire, but may cause an unacceptable level of thermal insulation to be applied to the cables. If this occurs, the cable rating needs to be reduced, as with any cable run in thermally insulating material.

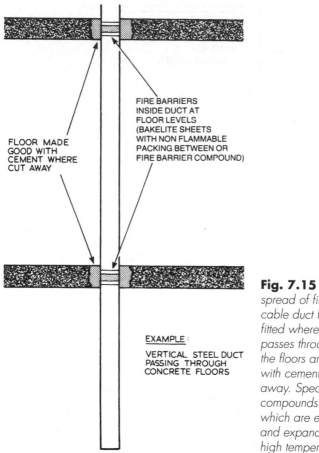

FIRE BARRIERS
INSIDE DUCT AT
FLOOR LEVELS
(BAKELITE SHEETS
WITH NON FLAMMABLE
PACKING BETWEEN OR
FIRE BARRIER COMPOUND)

FLOOR MADE
GOOD WITH
CEMENT WHERE
CUT AWAY

EXAMPLE :

VERTICAL STEEL DUCT
PASSING THROUGH
CONCRETE FLOORS

Fig. 7.15 *Preventing the spread of fire. In vertical cable duct fire barriers are fitted where the trunking passes through floors and the floors are made good with cement where cut away. Special fire barrier compounds are available which are elastomeric based and expand if exposed to high temperatures*

The identification of conduits and cables in buildings   IEE Regulation 514-01-02 demands that wiring is marked so that it can be identified during inspection, testing, repair or alteration. Table 4.2 shows the colour identification to be used for conductors and where appropriate, to distinguish conduit from other services such as oil, water, or gas pipes, it should be coloured orange (Fig. 7.16). Where it is not possible or economical to paint the full length, it suggests that sections adjacent to junction boxes and control points should be painted in the approved colour. The length of these sections should not be less than 300 mm. The incoming supply and the distribution cables should also be marked to show the nature of the supply (whether a.c. or d.c.), the number of phases and voltage.

Cables in low temperature areas   Some PVC insulated cables can be operated in temperatures down to −30°C and −40°C and

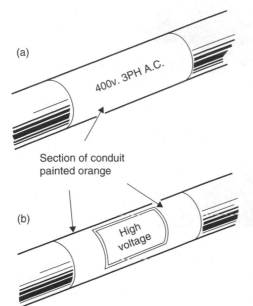

(a)

400v. 3PH A.C.

Section of conduit
painted orange

(b)

High
voltage

Fig. 7.16 (a) and (b)
Cables or conduit painted
orange – either the complete
length or 300 mm sections near
junctions and control points.
Identification letters are white
except for high voltage systems
and fire alarms where they are
black and red respectively

(c)

Fig. 7.16 (c) The
correct marking of
individual cable cores can
be greatly assisted by the
use of a power operated
portable marking machine
such as the one shown
here (Rejafix Ltd)

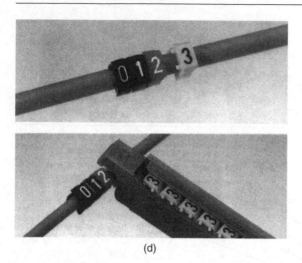

(d)

Fig. 7.16 *(d) In some situations, it is appropriate to use clip-on cable number-ing to identify individual cables. This system uses an applicator to apply individual numerals, and label holders for use on the larger size cables (Legrand Electric Ltd)*

manufacturers' data should be examined to check the minimum operating temperature of a particular cable type. However, cables should not be installed anywhere during periods when the tempera-ture is below 0°C, as the insulation is liable to crack if handled in very low temperatures.

Single-core cables Single-core cables, armoured with steel wire or tape shall not be used for a.c. circuits.

Bunching of outgoing and return cables If the outgoing and return cables of a 2-wire a.c. circuit, or all the phases and neutral of a 3-phase circuit, are enclosed in the same conduit or armoured cable excessive induction losses will not occur.

Single-core cables (without armour) enclosed in conduits or trunk-ing must be bunched so that the outgoing and return cables are enclosed in the same conduit or trunking. This must be accepted as a general rule for all a.c. circuits, and it must be ensured that no single conductor is surrounded by magnetic material, such as steel conduit, trunking or armouring. The reason is that any single-core cable car-rying alternating current induces a current in the surrounding metal, which tends to oppose the passage of the original current.

If the outgoing and return cables are enclosed in the same conduit or trunking, then the current in the outgoing and return cables, each carrying equal current, will cancel each other out as far as induction

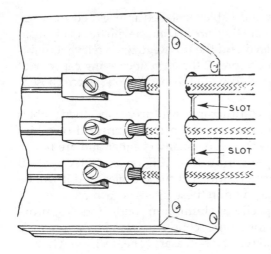

Fig. 7.17 *When single-core cables carrying heavy alternating currents pass through the metal casing of a switch, terminal box or similar equipment, they should, where possible, do so through a single hole; other-wise the space between the holes should be slotted to pre-vent circulating currents*

is concerned, and therefore no adverse effects will occur. If single-core cables are enclosed in separate metal conduit, trunking or metal armouring, the resultant induction losses could be very marked, and cause considerable voltage drop, and overheating of the cables and the enclosures.

A voltage drop of 90%, and considerable overheating, has been known when single-core cables, enclosed separately in magnetic metal have been connected to an a.c. supply.

Where it is essential that single core cables are used in a particular application, and the protection of conduit or trunking is required, consideration should be given to the use of non-metallic enclosures. A number of plastic conduit and trunking systems are now available.

There are occasions when the need to take precautions against induction is not observed. One example is when single-core cables enter busbar chambers, distribution boards or switchfuses; if single-core cables carrying alternating currents enter these through separate holes, circulating currents will be induced. In the past, manufacturers of switchgear and electric motors have provided three separate holes in the casing for 3-phase circuits. If it is impossible for all the cables to pass through one hole then the space between the holes should be slotted (Fig. 7.17).

Methods of installation

The IEE Regulations Appendix 4 and Table 4A give details of various types of installations and these affect the current-carrying capacity of the cables. Installation methods covered in IEE Table 4A include clipped direct, embedded in building materials, installed in conduit, trunking or on cable trays and cables installed in enclosed trenches. The installation of cables where they are in contact with thermally insulating materials is also covered and the IEE Regulation relevant to this is 523-04. As mentioned in Chapter 2 the current-carrying capacity of cables varies considerably according to the installation system chosen.

Cables with aluminium conductors   Multicore sheathed cables with aluminium conductors are sometimes used instead of cables with copper conductors, as they are usually cheaper and are not so heavy as cables with copper conductors. IEE Tables 4K1A to 4K4B and 4L1A to 4L4B give current ratings of these cables, and it will be noted that the smallest size given in these tables is 16 mm^2.

The current-carrying capacity of aluminium conductors is approximately 78% of the ratings for copper conductors, and therefore for a given current a larger cable will be necessary. The decision whether to use cables with copper or aluminium conductors will depend very much upon the market price of these two metals but of course other considerations must be taken into account, such as the fact that aluminium cables are generally of larger diameter than copper cables, and, as will be seen in the tables, the voltage drop is much greater (1.65 times that of copper).

The use of aluminium conductors presents some problems, but these can easily be overcome if the necessary precautions are taken. Aluminium, when exposed to air, quickly forms an oxide film which is a poor electrical conductor. If this film is allowed to remain it would set up a high resistance joint, and would cause overheating and eventually breakdown. There is also a risk, under damp conditions, of electrochemical action taking place between aluminium conductors and dissimilar metals. A further disadvantage is that the coefficient of expansion of aluminium is not the same as that for copper, and therefore terminations of aluminium conductors made to copper or brass terminals can give trouble if not properly made. Undoubtedly the best method of terminating aluminium conductors is to use crimping sockets made of tinned aluminium.

Before crimping, the conductors should be scraped to remove the oxide film, and then immediately smeared with Densal paste. Alternatively these conductors can be soldered to tinned copper sockets with the use of special fluxes and solders.

An alternative method is to connect a short length of copper conductor to the aluminium conductor by means of a crimped through connector, such as are used in aircraft wiring.

The IEE Regulations do not recognise cables with aluminium conductors smaller than 16 mm^2.

Segregation of cables

Where cables are associated with extra-low voltage, fire alarm and telecommunications circuits, as well as circuits operating at low voltage and connected directly to a mains supply system, precautions shall be taken to prevent electrical contact between the cables of the various types of circuit. This is covered by IEE Regulation 528-01 and this regulation refers to the voltage bands defined as follows:

Band I

Covers circuits operating at Extra Low Voltage (not exceeding 50 V a.c. or 120 V d.c.)(ELV) and also those where (a) protection against electric shock is provided by the value of the voltage in use and (b) voltage is limited for operational reasons such as with telecommunications, bell and alarm systems.

Band II

Covers circuits which supply household, commercial and industrial installations not exceeding 1000 V a.c. or 1500 V d.c.

No circuit of either Band I or Band II may be contained within the same wiring system as a circuit exceeding 1000 V a.c. (or 1500 V d.c.) unless one of the following arrangements is adopted:

(a) every cable is insulated for the highest voltage present;
(b) each conductor of a multicore cable is insulated for the highest voltage present in the cable;
(c) each conductor of a multicore cable is enclosed within an earthed metallic screen;
   or
(d) cables are installed in separate compartments of a duct or trunking system.

Similarly, no cable of a Band I circuit may be installed in the same wiring system as a cable of a Band II circuit unless placed in separate compartment of a duct or trunking although, as with the separation case above, some alternative methods are permitted. These include installations where:

(a) every cable is insulated for the highest voltage present;
(b) there is physical separation provided by a partition on a cable tray or ladder;
(c) they are installed in a separate conduit, duct or trunking.

Where cables enter a common box, circuits must be separated by partitions of fire-resisting material. Both metallic and non-metallic trunking systems are available with suitable barriers which ease installation and enable the requirements to be met.

The main object of these precautions is to ensure, in the case of fire, that alarm and emergency lighting cables are kept separate from other cables which might become damaged by the fire. BS 5266 deals with the segregation of these circuits and also gives details of other precautions which are necessary. Cables which are used to connect the battery chargers of self-contained emergency lighting luminaires to the normal mains circuits are not considered to be emergency lighting circuits.

Joints and connections between cables   Joints between cables should be avoided if possible, but if they are unavoidable they must be made either by means of suitable mechanical connectors or by soldered joints. In either case they must be mechanically and electrically sound and be readily accessible (IEE Regulations 131-12-01 and 526-04-01).

Electricity at Work Regulation 10 states that 'every joint and connection in a system shall be mechanically and electrically suitable for use'.

In some wiring systems, such as the MI and PVC sheathed wiring system, it is usual for the cables to be jointed where they branch off to lighting points and switches. These joints are made by means of specially designed joint boxes, or ceiling roses. In the conduit system it is usual for the cables to be looped from switch to switch and from light to light; joints should not be necessary and are to be discouraged.

Soldered joints   There are occasions when it become necessary to make joints between cables by means of soldering, although this method of jointing or terminating cables has now become almost obsolete due to the introduction of 'crimping', and a blowlamp, once an essential part of an electrician's equipment, is fast becoming antiquated.

Many years ago the practical test given to those aspiring to become electricians in the armed services or in civilian life was to demonstrate one's ability to make a soldered joint. As this method of jointing cables is very rarely taught nowadays it is as well to give some hints on the subject.

To make a good soldered joint the ends of the two pieces of cable should be bared of sheath and insulation for a length of approximately 50 mm. The conductors should then be carefully twisted

Fig. 7.18 *A high level cable route in an industrial building is illustrated in this view. Cable ladder is provided to carry distribution cables feeding switchboards remote from the site substation. Cable basket, which in this case has three separate compartments, is to be used for data cabling and cable tray is provided for final circuits feeding lighting and machinery (W.T. Parker Ltd)*

Fig. 7.19 *Another example of industrial wiring. Distribution cables feeding production machinery are carried on cable ladder. Cable basket is provided for the data cabling linking this satellite location to the main production control computers (W.T. Parker Ltd)*

Fig. 7.20   *Trunking and cable ladder are installed in this warehouse, preparatory to cabling being installed. Using the most appropriate system for the prevailing conditions enables a neat installation to be provided (W.T. Parker Ltd)*

together. The joint should be warmed and tested with solder until the solder tends to flow. A non-corrosive flux should then be applied and the solder allowed to flow freely over the whole of the joint.

After cooling the finished joint should be insulated by a suitable insulating tape or shrink-on sleeve.

The use of connectors   Small cables, not larger than 6 mm^2, may be connected by means of a fixed connecting block with grub screws. Larger cables should be connected with substantial mechanical clamps (not grub screws) and the ends of the cables should preferably be fitted with cable lugs. Lug terminals must be large enough to contain all the strands of the conductor and should be connected together with bolts and nuts or bolted to connecting studs mounted on an insulated base. There are various other types of mechanical clamps available which are suitable for connecting large cables. The jointing of MI copper-sheathed cables requires special consideration, and the method is described in Fig. 7.21.

Crimping   It is now common practice to use crimping lugs for terminating all sizes of copper and aluminium conductors. Crimping is carried out with special 'crimping lugs' and special tools as shown in Fig. 7.22. This makes a very efficient joint, and the need for special

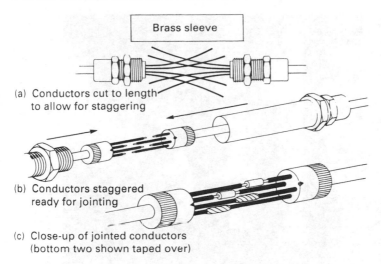

(a) Conductors cut to length
to allow for staggering

(b) Conductors staggered
ready for jointing

(c) Close-up of jointed conductors
(bottom two shown taped over)

Fig. 7.21 *Jointing of MI cables. The diagram shows the stages in the joint assembly. The cable ends are prepared as for a normal MI cable termination as shown in Chapter 13. A brass sleeve is provided to link two standard compression glands. The joints in the individual cores are staggered, and the connections crimped or soldered using a ferrule, and wrapped in tape. The joint is completed by sliding on the brass sleeve, and securing the nuts. If damp or hazardous conditions apply, the threads must be suitably sealed (BICC Pyrotenax)*

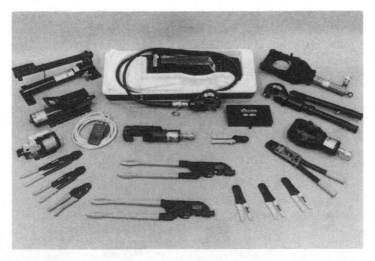

Fig. 7.22 *Cable crimping is commonly employed for cable terminations. The illustration shows a variety of crimping tools of both the mechanical and hydraulic types. (BICC Components Ltd)*

Fig. 7.22 *(Continued)*

solders and fluxes is eliminated. Crimping has now almost entirely replaced the old method of soldered joints.

However, to obtain satisfactory joints it is important to use the correct crimping tool and lug for the size of cable being jointed. The crimping tools must also be kept in good condition and the dies inspected periodically as any undue wear will result in unsatisfactory joints. This is particularly important where the connections will be subject to vibration, which can occur in a variety of industrial applications. British Standard BS 7609 offers much useful advice on crimp and cable preparation, and the procedures to be used to obtain satisfactory joints.

8 Conduit systems

A conduit is defined as a tube or channel. In electrical installation work 'conduit' refers to metal or plastic tubing. The most common forms of conduit used for electrical installations are steel conduits made to BS 31 or BS 4568, or uPVC plastic conduits. Non-ferrous metallic conduits mainly in copper and aluminium were formerly used in special installations but have now been virtually replaced by either the steel or plastic forms. Light gauge unscrewed steel conduit systems which were at one time used extensively in domestic installations, are now used only for such applications as intruder alarm systems.

The screwed steel conduit system

The steel conduit system is divided into two classes, class A which is plain, unscrewed, and class B which is screwed.

Class B screwed steel conduit is made in Imperial size (BS 31) and in metric sizes (BS 4568) (Table 8.1). These standards also give details of the threads for conduit and include details of standard conduit fittings, saddles, clips and other accessories. Practically all conduit fittings are made to these standards.

The screwed steel conduit system is undoubtedly the most popular for permanent wiring installations, especially for modern commercial and industrial buildings. Its advantages are that it affords the conductors good mechanical protection, permits easy rewiring when necessary, minimises fire risks, and presents a pleasing appearance if properly installed. Correct installation is important, and the general appearance of a conduit system reflects the degree of skill of the person who erected it.

The disadvantages are that it is expensive compared with some systems, is difficult to install under wood floors in houses and flats, and is liable to corrosion when subjected to acid, alkali and other fumes. Moreover, under certain conditions, moisture due to condensation may form inside the conduit.

Table 8.1   *Conduit dimensions*

Nearest Imperial size (in)	Metric size (mm)	Thickness of wall		Pitch of thread (mm)
		Heavy gauge (mm)	*Light gauge* (mm)	
⅝	16	1.6	1.0	1.5
¾	20	1.6	1.0	1.5
1	25	1.8	1.2	1.5
1¼	32	1.8	1.2	1.5

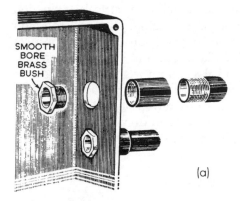

(a)

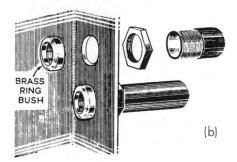

(b)

Fig. 8.1   *Two methods of fixing screwed conduit at clearance entries in metal casings or boxes. (a) With smooth bore bush and coupling; (b) with brass ring bush and backnut*

Types of screwed steel conduit

Class B screwed steel conduit is either solid drawn or welded, and each of these types and their fittings are supplied with alternative finishes as follows:

(1) Black enamel for internal use in dry situations.
(2) Silver grey finish for internal use in dry situations where this light finish will match decorations, or will provide a suitable base for light paint.

(3) Hot galvanised or sherardised for external use, or for internal use where subjected to dampness or steam.

The internal bore of all conduits must be smooth and free from burrs.

Solid drawn is much more expensive than welded conduit, and for this reason its use is generally restricted to gas-tight and explosion-proof installation work. Welded screwed conduit is, therefore, generally used for most installation work.

Cables in conduits

The types of cables which may be installed in conduits are PVC single-core insulated, butyl or silicone rubber insulated, with copper or aluminium conductors. PVC insulated and sheathed cables are sometimes installed in conduits when the extra insulation and protection is desirable.

The metric cables are rated at 600/1000 V, and these replace the old 440 V and 660 V grade. The metric cables are smaller in overall diameter than the Imperial sizes, due to a reduction in the thickness of the insulation, and, a larger number of cables may be drawn into a given size of conduit, than was permissible for the Imperial sizes.

One word of warning is necessary; cables up to $2.5\,mm^2$ now have solid conductors, and it has been found that these are not so easily drawn into conduit as are the stranded cables (i.e. the old 3/.029 to 7/.029), and it may be found that in practice it will be difficult to install the maximum number allowed by the Regulations.

This does not apply to butyl insulated cables which will be supplied as seven strand conductors for 1.5 and $2.5\,mm^2$ cables. It is, however, possible to install eight $2.5\,mm^2$ cables in 20 mm conduit and therefore two ring circuits can now be accommodated in the equivalent to the old ¾ in conduit. The 16 mm conduit, which is slightly larger than the old ⅝ in conduit, will accommodate six $1.5\,mm^2$ PVC insulated cables, but this size of conduit is rarely used in practice.

Choice of the correct size of conduit

It has already been explained how to determine the correct size of cables for any given load, and, having decided upon the type and size of cables to be used, it will then be necessary to select the appropriate size of conduit to accommodate these cables.

The number of cables which may be drawn into any conduit must be such that it allows easy drawing in, and in no circumstances may it be in excess of the maximum given in Table 3.5B. It will be noted that this table does not provide for cable larger than $10\,mm^2$.

Table 8.2 Spacing of supports for conduits (Extract from IEE On-site guide, Table 4C)

Nominal size of conduit (mm)	Maximum distance between supports							
	Rigid metal		Rigid insulating				Pliable	
	Horizontal (m)	Vertical (m)	Horizontal (m)	Vertical (m)			Horizontal (m)	Vertical (m)
Not exceeding 16	0.75	1.0	0.75	1.0			0.3	0.5
Exceeding 16 and not exceeding 25	1.75	2.0	1.5	1.75			0.4	0.6
Exceeding 25 and not exceeding 40	2.0	2.25	1.75	2.0			0.6	0.8
Exceeding 40	2.25	2.5	2.0	2.0			0.8	1.0

Note: A flexible conduit is not normally required to be supported in its run.

Fig. 8.2 *Conduit should be cut using a hack-saw and after cutting burrs should be removed using a file or reamer*

It must always be remembered that, as the number of cables or circuits in a given conduit or trunking increase, the current-carrying capacities of the cables decrease. It may therefore be advisable not to increase the size of the conduit or trunking in order to accommodate more cables, but to use two or more conduits.

Screwed copper conduit

Sometimes copper conduit is used. The advantage of copper conduit is that it resists corrosion and provides excellent continuity, but on the other hand the cost could prove to be prohibitive.

Copper conduit can be screwed in the same manner as steel conduit although the screwing of copper is more difficult than mild steel. Connections are generally made by soldering. Bronze junction boxes should preferably be used.

This system is comparatively expensive, but is used in buildings where long life and freedom from corrosion of the conduit and the cables is of first importance. For example, the House of Commons building is provided with copper conduits where the conduit system is buried in concrete floors and walls.

Installation of screwed conduit

The following refers to screwed steel conduits, although as already mentioned, copper conduits may be used instead of steel when desired.

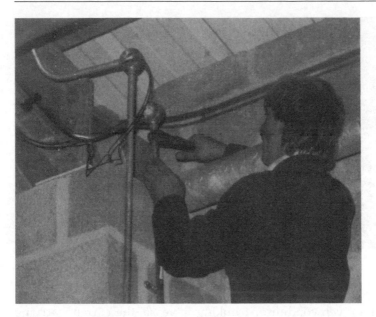

Fig. 8.3   *Installing a switch drop for a surface conduit system. A length of conduit, previously threaded at one end, is temporarily secured so that the length of run and position of fixing saddles can be marked off*

Screwed conduit is sometimes installed on the surface of walls, ceilings or trusses, or sometimes it is concealed in concrete or run under wood floors. There are therefore, two distinct methods of installing conduit, the surface system, and the concealed system. These two systems will be dealt with separately as far as the method of installation is concerned. Whichever system is adopted, a good deal of skill is required in order to produce a first class installation.

The surface system

Choice of runs   The first consideration is to choose the most suitable 'runs' for the conduits. When there are several conduits running in parallel, they must be arranged to avoid crossing at points where they take different directions. The routes should be chosen so as to keep the conduits as straight as possible, only deviating if the fixings are not good. The 'runs' should also be kept away from gas and water pipes and obstructions which might prove difficult to negotiate. Locations where they might become exposed to dampness or other adverse conditions should be avoided.

Fig. 8.5   Power supplies are generally available on site, and wall plug fixings may be drilled using power tools. These can be obtained with impact action for use on hard surfaces

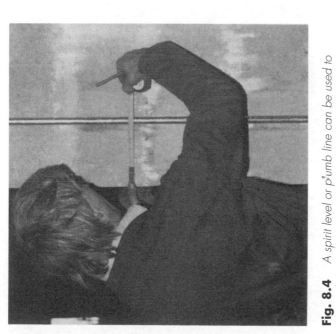

Fig. 8.4   A spirit level or plumb line can be used to ensure the conduit runs are vertical, or the position can be measured from the edge of finished blockwork as shown here

Fig. 8.7 The box is temporarily offered up and a spirit level is used to check the box for truth before marking off the fixing hole positions

Fig. 8.6 After cutting and threading the conduit, a coupler is screwed on ready to receive the box. The conduit saddle is dropped into position to enable the box to be located

Fig. 8.9   The conduit saddles are refitted and tightened and the box is screwed onto the wall

Fig. 8.8   After drilling the wall and inserting the wall plugs, the box is fixed and the brass 'bush' is inserted in position

Fig. 8.11   The cables are pulled in and after trimming to length, the socket is wired and fixed in position (William Steward & Co. Ltd)

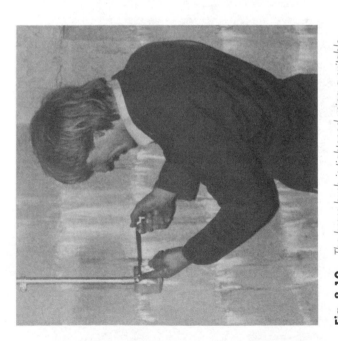

Fig. 8.10   The brass bush is tightened using a suitable spanner

Fig. 8.12   *The conduit bending machine incorporates a conduit vice which is useful for securing the conduit during thread cutting and assembly of fittings (William Steward & Co. Ltd)*

Conduit fittings   It is quite permissible to use manufactured bends, inspection tees and elbows, made in accordance with BS 31 and BS 4568. The use of these fittings should, however, be avoided for a first class surface installation. All bends should be made by setting the conduit, and where there are several conduits running in parallel which change direction it is necessary for these bends to be made so that the conduits follow each other symmetrically. This is not possible if manufactured bends are used.

In the author's opinion, the use of inspection elbows and tees is not good practice, as there is insufficient room for drawing in cables and, in addition, the installation presents a shoddy appearance.

A much more satisfactory method is to use round boxes to BS 31. These boxes have a much better appearance, provide plenty of room for drawing in cables, and can accommodate some slack cable which should be stowed in all draw-in points. For conduits up to 25 mm diameter, the small circular boxes should be used. These have an inside diameter of 60 mm. The larger circular conduit boxes are suitable for 32 mm diameter conduits.

Circular boxes are not suitable for conduits larger than 32 mm, and for these larger sizes rectangular boxes should be used. It would be impossible to draw large cables even into the larger type of circular box, as there would not be sufficient room to enable the final loop of the cable to be stowed into the box. Rectangular boxes are not covered by BS 31. They vary in size and some types are far too short for easy drawing in of cables, and they should therefore be selected to suit the size of cables to be installed.

The inspection sleeve is a very useful draw-in fitting, because its length permits the easy drawing in of cables and its restricted width

Fig. 8.13   *A selection of conduit fittings used with the screwed steel conduit system*

Fig. 8.14   *Conduit installation with the wiring in the process of being drawn in. The conduit is installed above the suspended ceiling and vertical conduits are run from Tee boxes (William Steward & Co. Ltd)*

enables conduits to be run in close proximity without the need to 'set' the conduits at draw-in points.

Where two or more conduits run in parallel, it is a good practice to provide at draw-in points an adaptable box which embraces all of the conduits. This presents a much better appearance than providing separate draw-in boxes and has the advantage of providing junctions in the conduit system which might prove useful if alterations have to be made at a later date.

The conduit system for each circuit should be erected complete before any cables are drawn in.

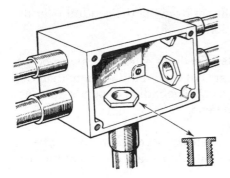

Fig. 8.15 *Where two or more conduits are run in parallel it is good practice to embrace all conduits with an adaptable box*

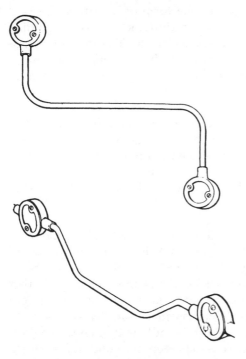

Fig. 8.16 *Cable must not be drawn round more than two right-angle bends or their equivalent. The four bends in the lower diagram are each at 45°, making a total of 180° in all*

An advantage of the conduit system is that the cables can be renewed or altered easily at any time. It is, therefore, necessary that all draw-in boxes should be readily accessible, and subsequently nothing should be fixed over or in front of them so as to render them inaccessible. The need for the conduit system to be complete for each circuit, before cables are drawn in, is to ensure that subsequent wiring can be carried out just as readily as the original; it prevents cables becoming damaged where they protrude from sharp ends of conduit, and avoids the possibility of drawing the conduit over the cables during the course of erection.

The radius of conduit bends   Facilities, such as draw-in boxes, must be provided so that cables are not drawn round more than two right-angle bends or their equivalent. The radius of bends must not be less than the standard normal bend.

Methods of fixing conduit   There are several methods of fixing conduit, and the one chosen generally depends upon what the conduit has to be fixed to.

Conduit clips   Conduit clips take the form of a half saddle, and have only one fixing lug. They are certainly more satisfactory than pipehooks, although the only reason for using clips instead of saddles is to save an additional fixing screw. They are not satisfactory if the conduit is subjected to any strain, and in any case their use gives a very parsimonious impression.

Ordinary saddles   Ordinary saddles are most commonly used and should be fixed by means of two screws – not nails. They provide a very secure fixing and should be spaced not more than 1.3 m apart. Conduit boxes to which luminaires are to be fixed should be drilled at the back and fixed, otherwise a saddle should be provided close to each side of the box.

Spacer bar saddles   Spacer bar saddles are ordinary saddles mounted on a spacing plate. These plates are 3 mm thick for conduits up to 40 mm diameter, and 5 mm for larger sizes of conduit. This spacing plate is approximately the same thickness as the sockets and other conduit fittings and, therefore, serves to keep the conduit straight where it leaves these fittings. When ordinary saddles are used the conduit is slightly distorted when the saddle is tightened. The real function of the spacer bar saddle is to prevent the conduit from making intimate contact with damp plaster and cement walls and ceilings which would result in corrosion of the conduit and discoloration of the decorations.

When conduit is fixed to concrete a high percentage of the installation time is spent in plugging for fixings, and the use of the spacerbar saddle which has only a one-hole fixing in its centre has an advantage over the ordinary saddle, in spite of the higher cost of the former. Some types of spacer bar saddles are provided with saddles having slots instead of holes. The idea is that the small fixing screws

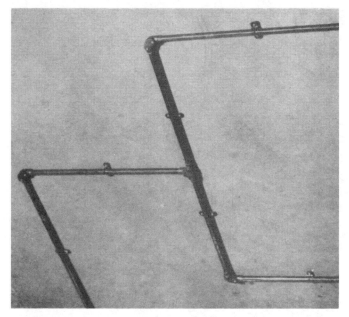

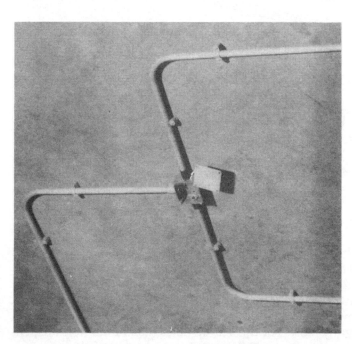

Fig. 8.17 *A neater and more workmanlike appearance is achieved by 'setting' the conduit and using a BS box rather than the use of elbows and tees as shown on the right (William Steward & Co. Ltd)*

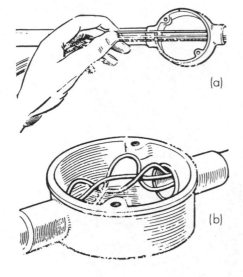

Fig. 8.18   *Cables must be fed in carefully to prevent them from crossing or becoming tangled as shown in (b). To thread not more than three cables, take them firmly between the fingers and thumb and keep them taut as shown in (a)*

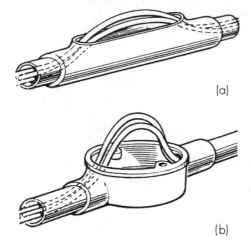

Fig. 8.19   *For drawing in large cables an inspection sleeve is often better than a round conduit box as with the latter it is often difficult to stow away the final loop. (a) Inspection sleeve (b) conduit box*

need only be loosened to enable the saddle to be removed, slipped over the conduit and replaced. This advantage is offset by the fact that when the saddle is fixed under tension there is a tendency for it to slip sideways clear of its fixing screws, and there is always a risk of this happening during the life of the installation if a screw should become slightly loose. For this reason holes rather than slots are generally more satisfactory in these saddles.

Table 8.3 *Minimum internal radii of bends in cables for fixed wiring (Extract from IEE On-site guide, Table 4E)*

Insulation	Finish	Overall diameter	Factor to be applied to overall diameter* of cable to determine minimum internal radius of bend
XLPE, PVC or rubber (circular, or circular stranded copper or aluminium conductors)	Non-armoured	Not exceeding 10 mm	3(2)†
		Exceeding 10 mm but not exceeding 25 mm	4(3)†
		Exceeding 25 mm	6
	Armoured	Any	6
XLPE, PVC or rubber (solid aluminium or shaped copper conductors)	Armoured or non-armoured	Any	8
Mineral	Copper or aluminium sheath with or without PVC covering	Any	6

*For flat cables the factor is to be applied to the major axis.
†The figure in brackets relates to single-core circular conductors of stranded construction installed in conduit, ducting or trunking.

When selecting the larger sizes of spacer bar saddles it is important to make sure that the slotted hole which accommodates the counter-sunk fixing screw is properly proportioned. These saddles are not covered by BS 31 and some of the fixing holes are made for no. 12 screws, although they are not countersunk deep enough to enable the top of the screw to be flush with the top of the spacing bar, and it therefore fouls the conduit.

Distance saddles These are designed to space conduits approxi-mately 10 mm from the wall or ceiling. Distance saddles are generally made of malleable cast iron. They are much more substantial than

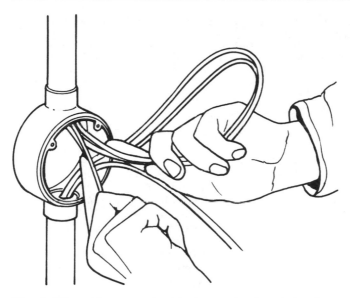

Fig. 8.20   *When several cables have to be threaded at the same time two hands are needed. The illustration shows the method of gripping the cable so as to guide into the conduit with the two forefingers*

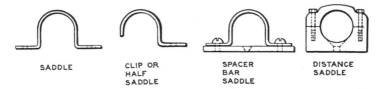

SADDLE   CLIP OR   SPACER   DISTANCE
                HALF   BAR   SADDLE
                SADDLE   SADDLE

Fig. 8.21   *Spacer bar saddles help to keep conduit straight and out of contact with damp plaster or cement. Distance saddles perform the same service, but are more robust*

Fig. 8.22   *A selection of standard clips designed for quick fitting of conduit to girders and other steelwork*

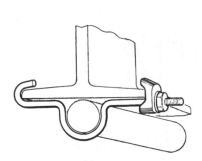

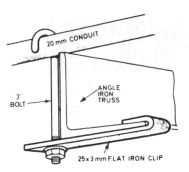

Fig. 8.23 *Girder clip is the best method of fixing conduit which runs along or across girders*

Fig. 8.24 *Method of fixing conduit to an angle iron truss with a flat iron clip and 'J' bolt*

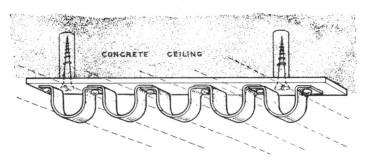

Fig. 8.25 *A 25 mm by 3 mm iron strap here is used to support five conduits on a concrete ceiling. It has two screw fixings*

other types of saddles, and as they space the conduit from the fixing surface they provide better protection against corrosion.

The use of this type of saddle eliminates the possibility of dust and dirt collecting behind and near the top of the conduit where it is generally inaccessible. For this reason distance saddles are usually specified for hospitals, kitchens, and other situations where dust traps must be avoided.

Multiple saddles Where two or more conduits follow the same route it is generally an advantage to use multiple saddles. The multiple saddles specified in BS 31 are not entirely satisfactory as they are designed for the conduits to touch each other. The proper method is for the conduits to be spaced so that when they enter conduit fittings there is no need to set the conduit.

Multiple spacer bar saddles can be purchased or they can be made up to suit a particular installation. An example is given in Fig. 8.25.

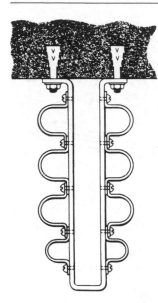

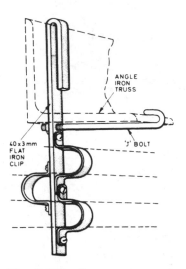

Fig. 8.26 *A U-section, fastened to a concrete ceiling with rag-bolts is used to carry a number of saddles of the required size*

Fig. 8.27 *Supporting several conduits from angle iron truss. This is a development of the method illustrated in Fig. 8.24*

Where several conduits have to be run on concrete, the use of multiple saddles saves a considerable amount of fixing time, as only two screws are required, and also ensures that all conduits are properly and evenly spaced.

Girder clips   Where conduits are run along or across girders, trusses or other steel framework, a number of methods of fixing may be used by the installer. A range of standard spring clips are available which can be quickly and easily fitted.

Other methods are also available including a range of bolt-on devices such as those shown in Figs. 8.23 and 8.24. If it is intended to run a number of conduits on a particular route and standard clips are not suitable, it may be advisable to make these to suit site conditions. Multiple girder clips can be made to take a number of conduits run in parallel, as shown in Fig. 8.27.

As an alternative to girder clips, multiple saddles can be welded to steelwork, or the steelwork could be drilled. However, structural steelwork should never be drilled unless permission is obtained from the architect or structural engineer.

Fig. 8.28   *Conduit installed in a hospital ward. The fixing saddles and joint boxes can be clearly seen, and the services at this bed-head location will be concealed with finishing panels when the wiring is complete (William Steward & Co. Ltd)*

When conduits are suspended across trusses or steelwork there is a possibility of sagging, especially if luminaires are suspended from the conduit between the trusses. These conduits should either be of sufficient size to prevent sagging, or be supported between the trusses. They can sometimes be supported by iron rods from the roof above. If the trusses are spaced 3 m or more apart it is not very satisfactory to attempt to run any conduit across them, unless there is additional means of support. It is far better to take the extra trouble and run the conduit at roof level where a firm fixing may be found.

Avoidance of gas, water and other pipes   All conduits must be kept clear of gas and water pipes, either by spacing or insulation. They must also be kept clear of cables and pipes which feed telephones, bells and other services, unless these are wired to the same standard as lighting, heating or power circuits. One exemption to this is that conduits may be fitted to electrically operated gas valves, and the like, if they are constantly under electrically skilled supervision. Another is that conduits may make contact with water pipes if they are intentionally bonded to them. They must not make casual contact with water pipes.

If conduits have to be run near gas or water pipes and there is a risk of them making contact, they should be spaced apart with wood

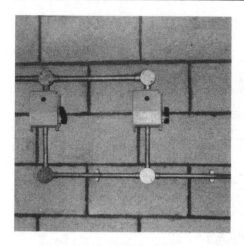

Fig. 8.29 *Surface conduit in use with isolating switches in an industrial location (William Steward & Co. Ltd)*

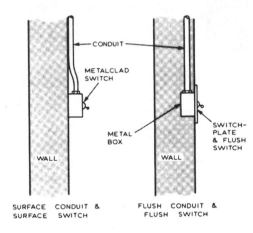

Fig. 8.30 *At switch positions conduit must terminate with metal box or other suitable enclosure. The illustration shows typical methods of terminating surface and concealed systems*

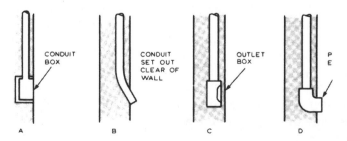

Fig. 8.31 *A box or suitable enclosure must be fitted at all outlet positions. Terminations as shown at B, C and D are not permitted*

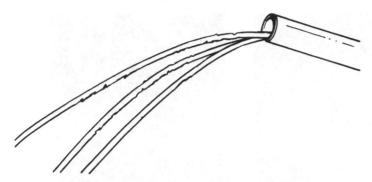

Fig. 8.32 *Sharp edges are usually formed inside conduit when it is cut and screwed. If these are not removed insulation of cable will be damaged*

or other insulating material. The reason for this precaution is that if the conduit system reaches a high potential due to defective cables in the conduit and an ineffective earth continuity, and this conduit makes casual contact with a gas or water pipe, either of which would be at earth potential, then arcing would take place between the conduit and the other pipe. This might result in puncturing the gas pipe and igniting the gas. If the gas or water pipe is of lead, this would very soon happen.

Protection of conduits

Although heavy gauge conduit affords excellent mechanical protection to the cables it encloses, it is possible for the conduit itself to become damaged if struck by heavy objects.

Such damage is liable to occur in workshops where the conduit is fixed near the floor level and may be struck by trolleys or heavy equipment being moved or slung into position. Protection can be afforded by threading a water pipe over the conduit during erection, or by screening it with sheet steel or channel iron. Another method of protection is, of course, to fix the conduit behind the surface of the wall.

Termination of conduit at switch positions

At switch positions the conduit *must* terminate with a metal box, or into an accessory or recess lined with incombustible materials.

Fig. 8.33   *Conduit is cut to length preferably by hacksaw, gripped in a vice and threaded*

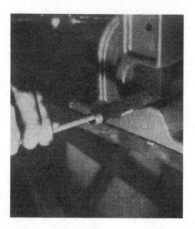

Fig. 8.34   *After cutting and screwing, burrs are removed with a round file or a reamer*

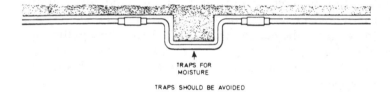

TRAPS FOR
MOISTURE

TRAPS SHOULD BE AVOIDED

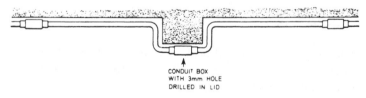

CONDUIT BOX
WITH 3mm HOLE
DRILLED IN LID

THIS METHOD AVOIDS A TRAP AND PERMITS EASIER
DRAWING IN OF CABLES

Fig. 8.35   *When there is a danger of condensation forming inside conduit (e.g. where there may be changes of temperature) holes should be drilled at the lowest points of the conduit system or, alternatively, conduit boxes with drainage holes should be fitted*

Termination of conduit at other than switch positions

Where conduit terminates at ceiling or wall points other than at switch positions, it must terminate with a metal box, or recess, or a block of incombustible material.

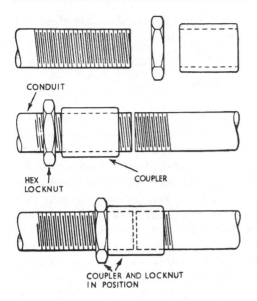

CONDUIT

HEX
LOCKNUT

COUPLER

COUPLER AND LOCKNUT
IN POSITION

Fig. 8.36 *For connecting two lengths of conduit, neither of which can be turned. The method of using the coupler and locknut may be clearly followed*

Removal of burrs from ends of conduit

When steel conduit is cut by a hacksaw, a burr is formed upon the inner bore of the conduit. If this burr were not removed it would cause considerable damage to the insulation of the cables drawn into the conduit. Fig. 8.32 shows what may happen to PVC cables when burrs are not removed. Ends of lengths of conduit should be free from burrs, and where they terminate at boxes, trunking and accessories not fitted with spout entries, should be treated so as to eliminate damage to cables.

Conduit installed in damp conditions

If conduits are installed externally, or in damp situations, they should either be galvanised, sherardised, or be made of copper, and all clips and fixings (including fixing screws) shall be of corrosion-resisting material (see IEE Regulation 522-03-01).

In these situations precautions must be taken to prevent moisture forming inside the conduit due to condensation. This is most likely to occur if the conduit passes from the outside to the inside of a building, or where there is a variation of temperature along the conduit route. In all positions where moisture may collect, holes should be drilled at the lowest point to allow any moisture to drain away. Drainage outlets should be provided where condensed water might otherwise collect. Whenever possible conduit runs should be designed so as to avoid traps for moisture.

Continuity of the conduit system

A screwed conduit system must be mechanically and electrically continuous across all joints so that the electrical resistance of the conduit, together with the resistance of the earthing lead, measured from the earth electrode to any position in the conduit system shall be sufficiently low so that the earth fault current operates the protective device (see Chapter 16). To achieve this it is necessary to ensure that all conduit connections are tight, and that the enamel is removed from adaptable boxes and other conduit fittings where screwed entries are not provided. For example some consumer control units are finished in a very pleasant stove enamel finish. They are provided with 'knock-outs' for conduit entry, but if the conduits are connected to these units it is necessary to remove the enamel at the position of the conduit entry if continuity is to be ensured. To guarantee the continuity of the protective conductor throughout the life of the installation, it is now common practice to draw a separate circuit protective conductor into the conduit for each circuit in the conduit.

Conduits must always be taken direct into distribution boards, switchfuses, switches, isolators, starters, motor terminal boxes, etc., and must be 'electrically and mechanically continuous throughout'.

Conduits must not be terminated with a bush and unprotected cables taken into switchgear and other equipment. The switchgear must be connected mechanically either with solid conduits, or with flexible metallic or plastic conduits.

Flexible conduit

For final connections to motors, it is usual to use pliable plastic conduit or flexible metallic conduit so as to provide for the movement of the motor if fixed on slide rails. It also prevents any noise or vibration being transmitted from the motor, or the machine to which it may be coupled, to other parts of the building through the conduit system (Fig. 8.37).

This flexible conduit should preferably be of the watertight pattern and should be connected to the conduit by means of brass adaptors. These adaptors are made to screw on to the flexible tubing and also into the conduit. It is good practice to braze the adaptor to the

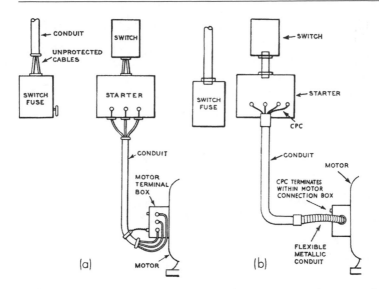

Fig. 8.37 *Wrong and right methods of terminating conduit at switch and starter. (a) Shows the wrong method, which is frequently adopted because proper conduit outlets are not always provided on starters and motors. The lengths of unprotected cable are subject to mechanical damage which may lead to electrical breakdown. (b) Illustrates the right method. Conduit is either taken direct into the equipment or terminated with flexible metallic conduit and a suitable CPC*

metallic tubing, otherwise it is likely to become detached and expose the cables to mechanical damage.

The use of flexible metallic tubing which is covered with a PVC sleeving is recommended, as this outer protection prevents oil from causing damage to the rubber insertion in the joints of the tubing. An alternative is to use pliable plastic conduit, and, in either case, an appropriate CPC must be provided.

Surface conduits feeding luminaires and clocks

Surface conduit run to feed wall or ceiling luminaires which are fixed direct to the wall or ceiling often presents a problem. If, for example, the conduit is run to feed an electric clock it means that the clock must either be spaced from the wall by a large wood block or other means, or the conduit must be 'set' into the wall so as to permit the clock to be fitted direct on to the face of the wall. It is advisable, if possible, to 'set' the conduit into the wall a short distance from the position of the clock (Fig. 8.38).

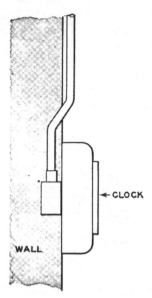

← CLOCK

WALL

Fig 8.38   *Surface conduit system when fitting or accessory must be flush on wall or ceiling*

Drawing cables into conduits

Cables must not be drawn into conduits until the conduit system for the circuit concerned is complete, except for prefabricated flexible conduit systems which are not wired *in situ*.

When drawing in cables they must first of all be run off the reels or drums, or the reels must be arranged to revolve freely, otherwise if the cables are allowed to spiral off the reels they will become twisted, and this would cause damage to the insulation (Fig. 8.39). If only a limited quantity of cable is to be used it may be more convenient to dispense it direct from one of the boxed reels which are now on the market.

If a number of cables are being drawn into conduit at the same time, the cable reels should be arranged on a stand or support so as to allow them to revolve freely (Fig. 8.40).

In new buildings and in damp situations the cable should not be drawn into conduits until it has been made certain that the interiors of the conduits are dry and free from moisture. If in doubt, a draw wire with a swab at the end should be drawn through the conduit so as to remove any moisture that may have accumulated due to exposure or building operations.

It is usual to commence drawing in cables from a mid-point in the conduit system so as to minimise the length of cable which has to be

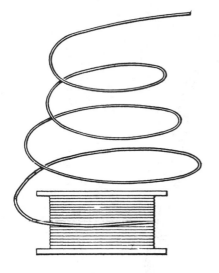

Fig. 8.39   *Cable must not be allowed to spiral off reels or it will become twisted and the insulation damaged. It should be run off by a method similar to that shown in Fig. 8.40*

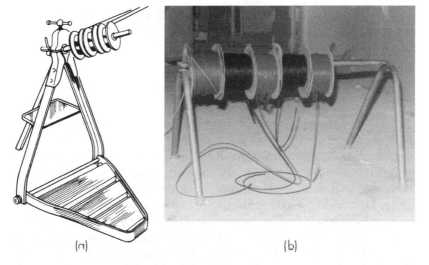

(a)   (b)

Fig. 8.40   *Running off cables from reels. (a) Illustrates a typical method used when only a few cables are involved. A short piece of conduit is gripped in a pipe vice. When many reels have to be handled it is best to use a special rack such as that shown in (b)*

drawn in. A draw-in tape should be used from one draw-in point to another and the ends of the cables attached. The ends of the cables must be bared for a distance of approximately 50 mm and threaded through a loop in the drawtape. When drawing in a number of cables

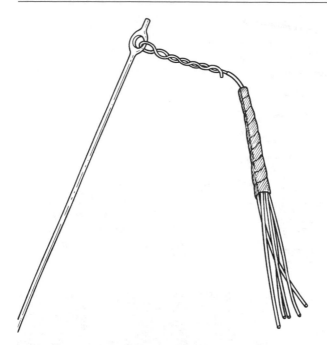

Fig. 8.41 *How to connect cable to draw tape*

they must be fed in very carefully at the delivery end whilst someone pulls them at the receiving end.

The cables should be fed into the conduit in such a manner as to prevent any cables crossing, and also to avoid them being pulled against the sides of the opening of the draw-in box. In hot weather or under hot conditions, the drawing-in can be assisted by using cable-pulling lubricant. Always leave some slack cable in all draw-in boxes and make sure that cables are fed into the conduit so as not to finish up with twisted cable at the draw-in point (see Fig. 8.18).

This operation needs care and there must be synchronisation between the person who is feeding and the person who is pulling. If in sight of each other this can be achieved by a movement of the head, and if within speaking distance by word of command given by the person feeding the cables. If the two persons are not within earshot, then the process is somewhat more difficult. A good plan is for the individual feeding the cables to give pre-arranged signals by tapping the conduit with a pair of pliers.

In some cases, it may be necessary for a third person to be stationed midway between the two positions to relay the necessary instructions from the person feeding to the person pulling. If cables are not drawn in carefully in this manner, they will almost certainly become crossed and this might result in the cables becoming jammed inside the conduit. In any case, it would prevent one or more cables being drawn out of the conduit should this become necessary.

The number of cables drawn into a particular size conduit should be such that no damage is caused to either the cables or to their enclosure during installation. It will be necessary, after deciding the number and size of cables to be placed in a particular conduit run, to determine the size of conduit to be used. Each cable and conduit size is allocated a factor and by summing the factors for all the cables to be run in a conduit route, it is an easy matter to look up the appropriate conduit size to use.

For example, if it is desired to run eight 2.5 mm^2 and six 4.0 mm^2 cables along a 4 m run of conduit with two bends, it is possible to determine the conduit size as follows:

From Table 3.5B, factors for 2.5 mm^2 and 4.0 mm^2 cables are 30 and 43 respectively.

$$8 \times 30 = 240$$
$$6 \times 43 = 258$$
$$\text{Total} = 498$$

Refer to Table 3.5A. As can be seen 25 mm diameter conduit with a factor of 388 would be too small; 32 mm diameter conduit with a factor of 692 will be suitable.

Looping in

When wiring an installation with PVC covered cable in conduit, joints are avoided as far as possible, and the looping-in system is normally adopted. In practice when wiring in conduit, the two lengths of cable forming the loop are threaded in separately and the junction is made at the switch, light, or other terminal.

Some practical hints

Apart from the electrician's ordinary tools, such as rule, hacksaw, hammer, screwdriver, pliers, etc., it is necessary to have stocks and dies, file or reamer, bending block or bending machine and a pipe vice. For 16 mm and 20 mm conduit, the small stocks are suitable, but for 25 mm and 32 mm, the medium stocks should be used. Although

25 mm dies are provided for the small stocks it is best to use the medium stocks for 25 mm conduit.

Conduit screwing machines

Conduit screwing machines are now on the market, including a motorised machine which offers certain advantages on a conduit installation where a considerable amount of large conduit is being installed.

Stocks and dies for screwing conduit should be clean, sharp and well lubricated, and should be rotated with a firm and steady movement. To get the best results stocks and dies should be of the self-clearing pattern to prevent the soft swarf from clogging the chasers. Dies should have a considerable top rake on the cutting face and the back face should be radial to allow cutting to take place when backing out.

For machine screwing a top rake angle of 25° to 35° is recommended, giving a cutting angle of 55° to 65°. Side rake is very desirable since it helps to clear the swarf, and is best obtained by grinding the nose of the chasers to produce a left-hand spiral cutting tool which will direct cuttings forward and out of the die. A side rake of 15° to 20° extending back about three teeth from the nose of the chaser is recommended and a 30° chamfer should be ground on the first two teeth of the cutters.

Lard oil or graphite grease is best for conduit dies. These should always be kept lubricated and never used dry. Worn dies and guides should always be replaced when showing signs of wear, otherwise the workmanship will suffer as a result of bad threads.

Conduit cutting

Conduit should be cut with a hacksaw in preference to a pipe cutter, as the latter tends to cause a burr inside the conduit.

In any case, the ends of all conduit must be carefully reamered inside the bore with a file, or reamer, to be certain that no sharp edges are left which might cause damage to the cables when they are being drawn in. This reamering should be carried out after the threading has been completed.

Checking conduit for obstructions

When the length of conduit has been removed from the pipe vice, it is a good idea to look through the bore to ensure that there are no obstructions. Some foreign object, such as a stone, may have entered the conduit during storage (especially if stored on end) or welding metal may, in rare cases, have become deposited inside the conduit. If such obstructions are not detected before the installation of the conduit, considerable trouble may be experienced when the cables are drawn in.

Bending conduit

For bending or 'setting' conduit, either a bending block or bending machine should be used.

The best bending blocks are made of ash as this wood is not too hard and not too soft. A convenient size is 100 mm by 50 mm, about 1.3 m long with holes drilled within 150 mm from each end for the size of conduit to be used. The bottom edge of each hole should be bevelled so that the conduit is not pulled against a sharp edge. Practice is necessary to acquire skill in using a bending block, and this is probably the most skilled part of an electrician's craft. Nowadays it is usual to use a bending machine for all sizes of conduit which enable bends and sets to be made without the risk of kinks or flattening of the conduit. These machines are also necessary when very sharp bends have to be made in 16 mm and 20 mm conduits, and especially for copper conduits.

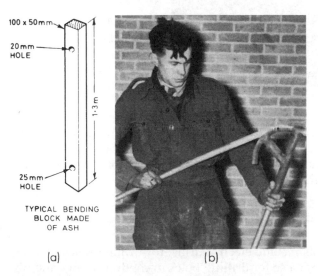

(a)   (b)

Fig. 8.42   *(a) Small conduit may be bent using a bending block. (b) Shows a simple tool for bending small-sized conduit*

Fig. 8.43 Bending machine for medium-sized conduit

Fig. 8.44 Detail of the conduit bending machine showing the former over which the conduit is placed, and the roller (Hilmor Ltd)

Fillers, such as sand and compounds of low melting point are not generally recommended because of the difficulty of handling, and spring fillers may be difficult to withdraw. Means of providing external support are therefore adopted.

Drilling and cutting

Other useful tools include electric drills, for drilling holes in structural ironwork and also for drilling holes in conduit fittings and distribution boards. Suitable hole cutters for cutting holes can be used in electric drills. Electric hammers save a considerable amount of time for wall plugging, cutting holes through walls and floors and also for cutting chases in walls for conduit. Safety eye shields must be worn. As a final word of advice, do not forget to make good the holes after the conduit has been erected.

The concealed screwed conduit system

Screwed metal conduit is particularly suitable for concealed wiring. The conduits can be installed during building operations and can be safely buried in floors and walls in such a manner that the cables can be drawn in at any time after the completion of the building. Whether the floors or walls are constructed of wood, brick, hollow tiles or solid concrete, the conduit system, if properly installed, can be relied upon adequately to protect the cables and allows them to be replaced at any time if desired.

Most modern buildings, including blocks of flats, are constructed with solid floors and solid walls and it is necessary for the conduit (if concealed) to be erected during the construction of the building. In other types of buildings where there are wooden joists and plaster ceilings, conduit will have to be run between and across the joists.

Running conduit in wooden floors

Where conduit is run across the joists, they will have to be slotted to enable the conduit to be kept below the level of the floorboards. When slots are cut in wooden joists they must be kept within 0.1 to 0.2 times the span of the joist, and the slots should not be deeper than 0.15 times the joist depth (Fig. 8.45).

The slots should be arranged so as to be in the centre of any floorboards, if they are near the edge there is the possibility of nails being driven through the conduit.

'Traps' should be left at the position of all junction boxes. These traps should consist of a short length of floorboard, screwed down and suitably marked.

Running conduit in solid floors

Where there are solid floors the conduit needs to be arranged so that cables can be drawn in through ceiling or wall points. This method is known as the 'looping-in system', and it is shown in

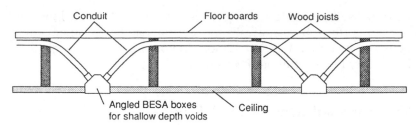

Fig. 8.45   *Running conduit in wood floors to feed lighting points. The angled BESA boxes allow for a shallow bend in the conduit, which is fixed to the box with couplings and male brass bushes. The box can, if required, be terminated above the ceiling and extension rings used to make the outlet flush with the ceiling. This allows wiring without disturbing the floor. The slots in the joists should be in a zone 0.1 to 0.2 times the span of the joist with a maximum depth of 0.15 times the joist depth (Trevor Marks)*

Fig. 8.46   *Wiring a commercial installation with PVC-insulated cable in concealed steel conduit (William Steward & Co. Ltd)*

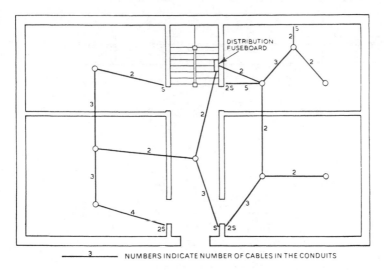

Fig. 8.47 *Typical arrangement of concealed conduits feeding lighting points by looping the conduit into the back of outlet boxes*

Figs. 8.47 and 8.48. Conduit boxes are provided with holes at the back to enable the conduit to be looped from one box to another. These boxes are made with two, three or four holes so that it is possible also to tee off to switches and adjacent ceiling or wall points.

The correct method of fixing the conduit to these boxes is to fit a socket outside the back of the box, and a hexagon brass bush inside. The bush should be firmly tightened with a special box spanner, and consultants very often specify that copper compression washers be fitted between the bush and the box. If these joints in the conduit system are not absolutely tight there might be difficulty in obtaining a satisfactory continuity test on completion. Satisfactory continuity is essential.

There are many types of solid or semi-solid floors and some of these do not lend themselves to the installation of concealed conduit. Some types are so shallow that a very sharp set has to be made in the conduit after it leaves the socket, and it is best to use a bending machine with a special small former made for these types of floors.

If the floors are of reinforced concrete, it may be necessary to erect the conduit system on the shuttering and to secure it in position before the concrete is poured. If not securely fixed, it may move out of position or lift and then, when the concrete is set, it will be too late to rectify matters.

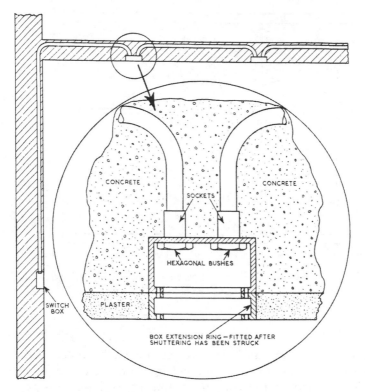

Fig. 8.48   *Details of conduit box and method of fastening conduit. A socket is fixed outside the back of the box and a brass hexagonal bush inside. The bush should be firmly tightened, otherwise there will be difficulty in obtaining a satisfactory continuity test on completion*

Wherever conduit is to be buried by concrete, special care must be taken to ensure that all joints are tight, otherwise liquid cement may enter the conduit and form a solid block inside.

Preferably the joints should be painted with a bitumastic paint, and the conduit itself should also be painted where the enamel has been removed during threading or setting.

Sometimes the conduits can be run in chases cut into concrete floors; these should be arranged so as to avoid traps in the conduit where condensation may collect and damage the cables.

Conduit runs to outlets in walls

Sockets near skirting level should preferably be fed from the floor above rather than the floor below, because in the latter case it would be difficult to avoid traps in the conduit (Fig. 8.49).

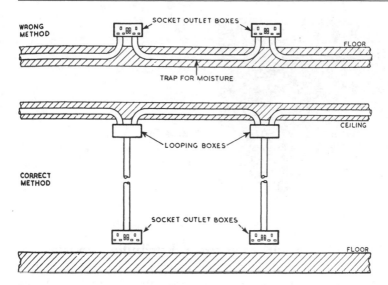

Fig. 8.49   *Right and wrong methods of feeding sockets near skirting level. If the sockets are fed from the floor below, it is difficult to avoid a trap for moisture*

When the conduit is run to switch and other positions in walls it is usually run in a chase in the wall. These chases must be deep enough to allow at least 10 mm of cement and plaster covering, otherwise rust from the conduit may come through to the surface. Conduits buried in plaster should be given a coat of protective paint, or should be galvanised if the extra cost is justified.

Make sure that the plaster is finished neatly round the outside edges of flush switch and socket boxes, otherwise the cover plates may not conceal any deficiencies in the plaster finish. When installing flush boxes before plastering, it is advisable to stuff the boxes with paper to prevent their being filled with plaster.

Ceiling points

At ceiling points the conduit boxes will be flush with the finish of the concrete ceiling. If the ceiling is to have a plaster rendering, this will leave the front of the boxes recessed above the plaster finish. To over-come this it is possible to purchase extension rings for standard con-duit boxes (Fig. 8.48).

At the position of ceiling points it is usual to provide a standard round conduit box, with an earth terminal, but any metal box or incombustible enclosure may be used, although an earth terminal must be provided.

Fig. 8.50   *A metal conduit box with an extension piece for use where the depth has to be increased as building work progresses. See Fig. 8.48*

Running sunk conduits to surface distribution boards

Where surface mounted distribution boards are used with a sunk conduit, the problem arises as to the best method of terminating flush conduits into the surface boards. One method is to 'set' the conduits out to the required distance into the surface boards (Fig. 8.50), but this presents a very unsightly appearance, and is not recommended.

A better method is to fit a flush adaptable box in the wall behind the distribution board, and to take the flush conduits directly into it. Holes can be drilled in the back of the distribution board, and bushed. Spare holes should be provided for future conduits. Alternatively, an adaptable box can be fitted at the top of the distribution board, partly sunk into the wall to receive the flush conduits, and partly on the surface to bolt on the top of the distribution board. Distribution boards must be bonded to the adaptable boxes.

Before wiring sunk conduit

Before wiring, the conduits for each circuit must be erected complete. Not only should they be complete but they must be clean and dry inside otherwise the cables may suffer damage. No attempt should be made to wire conduits which are buried in cement until the building has dried out and then the conduits should be swabbed to remove any moisture or obstructions which may have entered them.

The light gauge unscrewed conduit system

The light gauge conduit system consists of Class A metal conduits to BS 31, which are made of a light gauge strip and the seams are either

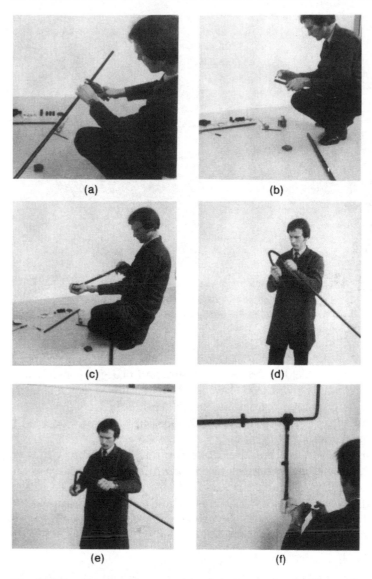

Fig. 8.51 *Stages in the assembly of plastic conduit. (a) The conduit can easily be cut to length with the special cutters obtainable for the purpose. (b, c) Jointing is carried out by the use of solvent adhesives, and care must be taken in their application to avoid blocking small-sized conduits. (d, e) Bending can be by hand and a bending spring is used to retain the shape of the bore. The conduit is bent through almost 180° and then returned to the right angle generally needed for bends. (f) The finished installation presents a similar appearance to steel systems (MK Ltd)*

Fig. 8.52   *Drawing cables into a vertical conduit. The operation needs care, and there must be synchronisation between the person feeding and the person who is pulling. The cables should be fed in so as to prevent any cables crossing or becoming twisted at the draw-in point (William Steward & Co. Ltd)*

butted, brazed or welded. The walls of this conduit are not of sufficient thickness to allow them to be threaded. Instead of screwed sockets and fittings grip type fittings are used. Although used during the 1930s and 1940s for domestic installations the system is now only used for intruder alarm systems.

The practice of threading the cables through the conduit and then connecting the elbows and tees to the conduit was customary in the days of speculative building when first cost was the only consideration. After the completion of the building it was quite impossible to withdraw or renew any of the cables when trouble developed, as it frequently did. This method of wiring is now contrary to IEE Regulations.

Insulated conduit system

Non-metallic conduits are now being increasingly used for all types of installation work, both for commercial and house wiring. The PVC rigid conduit is made in all sizes from 16 mm to 50 mm in external diameter, and there are various types of conduit fittings, including boxes available for use with this conduit.

Fig. 8.53   *Insulated conduit system installed in merchant ship (MK Ltd)*

Fig. 8.54 shows a range of typical components made of a plastic material, and fitted with special sockets which enable the conduit to be assembled neatly. Solvent adhesives are available for jointing though a suitable number of sliding joints must be left to allow for expansion.

The advantage of the insulated conduit system is that it can be installed much more quickly than steel conduit, it is non-corrosive, impervious to most chemicals, weatherproof, and it will not support combustion. The disadvantages are that it is not suitable for temperatures below −5°C, or above 60°C, and where luminaires are suspended from PVC conduit boxes, precautions must be taken to ensure that the heat from the lamp does not result in the PVC box reaching a temperature exceeding 60°C.

For surface installations it is recommended that saddles be fitted at intervals of 800 mm for 16 mm diameter conduit, and intervals

Fig. 8.54   *A range of fittings for use with plastic conduit include the boxes, bends, T pieces, couplers and clips seen above (Marshall Tufflex Ltd)*

of 1600 mm to 2000 mm for larger sizes. The special sockets and saddles for this type of conduit must have provision to allow for longitudinal expansion which may take place with variations in ambient temperature.

It is of course necessary to provide a circuit protective conductor in all insulated conduit, and this must be connected to the earth terminal in all boxes for switches, sockets and luminaires. The only exception is in connection with Class 2 equipment, i.e. equipment having double insulation (see IEE Regulation 413-03-01). In this case a protective conductor must not be provided.

Flexible PVC conduits are also available, and these can be used with advantage where there are awkward bends, or under floorboards where rigid conduits would be difficult to install.

Installation of plastic conduit

Plastic conduits and fittings can be obtained from a number of different manufacturers and the techniques needed to install these are not difficult to apply. Care is however needed to assemble a neat installation and the points given below should be borne in mind. As with any other installation good workmanship and the use of good quality materials are essential.

It should be noted that the thermal expansion of plastic conduit is about six times that of steel, and so whenever surface installation of straight runs exceeding 6 m is to be employed, some arrangement

Fig. 8.55   A neat completed conduit installation, ready for wiring to be installed (William Steward & Co. Ltd)

must be made for expansion. The saddles used have clearance to allow the conduit to expand. Joints should be made with an expansion coupler which is attached with solvent cement to one of the lengths of tube, but allowed to move in the other.

Cutting the conduit can be carried out with a fine toothsaw or using the special tool shown in Fig. 8.51. As with steel conduit, it is necessary to remove any burrs and roughness at the end of the cut length.

Fig. 8.56   *Plastic conduit and fittings in this installation have been arranged with socket outlets for feeding lighting fittings. This work has been completed prior to the erection of the suspended ceiling (British Telecom)*

Fig. 8.57   *Non-metallic conduit showing some of the range of fittings available (Gilflex Ltd)*

Bending the small sizes of plastic conduit up to 25 mm diameter can be carried out cold. A bending spring is inserted so as to retain the cross-sectional shape of the tube. It is important to use the correct size of bending spring for the type of tube being employed. With cold bending, the tube should initially be bent to about double the required angle, and then returned to the angle required, as this reduces the tendency of the tube to return to its straight form. To bend larger sizes of tube, 32 mm diameter and above, judicious application of heat is needed. This may be applied by blowlamp, electric fire or boiling water. If a naked flame is used, extreme care must be taken to avoid overheating the conduit. Once warm, insert a bending spring and bend the tube round a suitable former. A bucket is suitable, but do not use a bending machine former, as this conducts away the heat too rapidly. The formed tube should as soon as possible be saddled after bending.

Joints are made using solvent adhesives which can be obtained specifically for the purpose. These adhesives are usually highly flammable and care is needed in handling and use. Good ventilation is essential, and it is important not to inhale any fumes given off. Clean and dry the components to be joined before commencing work. Avoid using excess solvent as this may block the conduit by forming a barrier across the inside especially when joining small size conduits. Using too little solvent may not make a waterproof joint. Experience will indicate the correct quantity of adhesive to use. The manufacturers' instructions for use of the solvent adhesive should be strictly followed. It is generally necessary to apply the adhesive to both surfaces to be joined, pushing the components together and holding them steady for about 15 s without moving to ensure the joint is set. Where expansion joints are needed the expansion collar should be solvent welded to one length of tube, but left free to slide on the other. If sealing is needed to waterproof the joint, use a special non-setting adhesive or grease. Threaded adaptors are available for use when it is required to make connections to screwed systems. These can be solvent welded to the plastic tube and screwed into the threaded fitting as required.

Drawing in cables is carried out by making use of a nylon draw-in tape. The smooth bore of the plastic tube aids the pulling in operation. Cable-pulling lubricant may be used to help the pulling in process. Capacities of plastic conduits may be calculated in a similar way to that used for steel systems. The method employed is as described for steel conduit and uses a 'unit system'. Each type of cable is allocated a factor, and corresponding factors are allocated for various sizes of conduit. Tables 3.5A and 3.5B give the factors applicable to cables and conduits. By means of this method, use of excess force is avoided, and it can be ensured that as cables are drawn into conduit damage to both cables and conduit is avoided.

The use of plastic conduit is suitable when cable runs require to be located in pre-cast concrete. As will be appreciated it is essential that sound joints are made so that when the concrete is cast, the conduit runs do not separate. This system is illustrated in use in Fig. 7.2.

9 Trunking systems

Metallic trunking

Trunking for industrial and commercial installations is often used in place of the larger sizes of conduit. It can be used with advantage in conjunction with 16 mm to 32 mm conduits, the trunking forming the background of framework of the system with conduits running from the trunking to lighting or socket outlet points. For example, in a large office building, trunking can be run above the suspended ceiling along the corridors to feed corridor points, and rooms on either side can be fed from this trunking by conduits.

In multistorey buildings trunking of suitable capacity, and with the necessary number of compartments, is usually provided and run vertically in the riser ducts and connected to distribution boards; it can also accommodate circuit wiring, control wiring, also cables feeding fire alarms, telephones, emergency lighting and other services associated with a modern building.

As explained in Chapter 7, cables feeding fire alarms and emergency circuits need to be segregated by fire-resisting barriers from those feeding low-voltage circuits (i.e. 50 V to 1000 V). It is usual for telecommunications companies to insist that their cables are completely segregated from all other wiring systems. It may therefore be necessary to install 3 or 4 compartment trunking to ensure that IEE Regulation 528-01 and the requirements for data and telecommunications circuits are complied with. Cables feeding emergency lighting and fire alarms must also be segregated so as to comply with the requirements of BS 5266. Additionally, segregation may be required to achieve electromagnetic compatibility requirements.

Many reputable manufacturers can now supply trunking ranging from 38 mm × 38 mm to 300 mm × 300 mm or even larger, with two or more compartments. They also provide all necessary accessories such as bends, tees, crossovers and bridges to segregate cables of different systems at junctions.

Table 9.1 *Spacing of supports for trunking (IEE On-site guide, Table 4D)*

Cross sectional area (mm²)	Maximum distance between supports			
	Metal		Insulating	
	Horizontal (m)	Vertical (m)	Horizontal (m)	Vertical (m)
exceeding 300 and not exceeding 700	0.75	1.0	0.5	0.5
exceeding 700 and not exceeding 1500	1.25	1.5	0.5	0.5
exceeding 1500 and not exceeding 2500	1.75	2.0	1.25	1.25
exceeding 2500 and not exceeding 5000	3.0	3.0	1.5	2.0
exceeding 5000	3.0	3.0	1.75	2.0

Trunking systems are more flexible than conduit systems. Extensions can readily be made during the life of the installation by punching a new hole in the trunking and running a conduit to a new point. Naturally, care is needed with the design of such an alteration as grouping of additional circuits may require the derating of cables to be re-assessed. However, it may be possible to implement the alteration without disturbing the existing wiring.

Trunking can be easily and quickly erected, and can be fitted to walls or suspended across trusses; where suspended it should be supported at each joint.

Where there are vertical runs of trunking pin racks should be fitted inside the trunking to support the weight of the cables, and to enable the cables to be secured during installation. These pin racks consist of steel pins, sheathed by an insulating material, mounted on a back-plate; they should be fitted at intervals of 5 m.

Where vertical trunking passes through floors it must be provided with internal fire barriers, which must consist of nonflammable materials, cut away to enable cables to pass through and made good after the installation of the cables.

When large cables are installed in trunking care must be taken to ensure that all bends are of sufficient radius to avoid damaging the cable (IEE Regulation 522-08-03). The IEE Guidance Notes give useful advice on this subject. These state, for example, that non-armoured PVC-insulated cables of an overall diameter greater than 25 mm shall not be so bent that the radius of the inside of the bend is less than six times the diameter of the cable. Trunking manufacturers provide bends and tees that enable this requirement to be satisfied.

Trunking can be used to accommodate PVC insulated cables that are too large to be drawn into conduit. Unless there are special reasons for using conduit, it will generally be found more economical to use trunking rather than conduit larger than 32 mm diameter.

Skirting trunking

Skirting trunking is now used extensively in commercial buildings, laboratories, hospitals and similar installations. It usually consists of a shallow steel trunking, approximately 44 mm deep with two or more compartments. One compartment is used for socket or lighting wiring, one for bell or internal telephone wiring, and very often a third compartment is reserved for telephone wiring, as these cables must be separated from all other wiring systems.

The trunking can be shaped to form the skirting, and is frequently fitted round the outer walls of a building where sockets, telephones, etc., are likely to be required. It can also be fitted on internal walls when its use can be justified. In order to cross the thresholds of doorways, and to interconnect isolated lengths of skirting trunking, conduits or floor trunking can be installed in the floor screed. Suitable bends and adaptors are made to connect between skirting and floor trunking.

Shallow flush-type socket-outlets (Fig. 9.1) can be obtained for fitting to the lid of skirting trunking and trunking manufacturers will punch suitable apertures for the reception of sockets.

It is often an advantage to fit the sockets or telephone outlets on short lengths of lid, which need not be disturbed when the remainder of the lid is removed for extensions.

Where trunking passes through partitions, short lengths of lids should be fitted as this enables the remainder of the lid to be removed without difficulty.

As with all types of steel trunking, there must be good continuity between all joints.

Regulations

The regulations governing wiring in conduit also apply to wiring in trunking, as far as applicable. All sections of trunking, bends, and other accessories must be effectively earthed in order to ensure that the conductivity of the trunking is such as to enable earth-fault current to flow to operate the fuse or earth-leakage circuit breaker protecting the circuit.

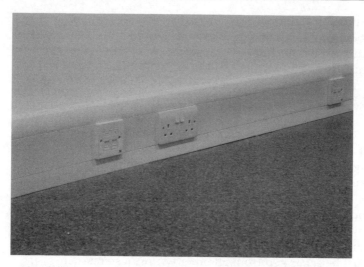

Fig. 9.1 *Multi compartment skirting trunking allows the segregation of different types of circuit. In this example 13-amp ring main socket outlets and telecommunications circuits are provided. (W.T. Parker Ltd)*

Trunking is usually supplied in 3 m lengths, although in some cases longer lengths are obtainable. If copper links are fitted these will generally ensure satisfactory earth continuity, but if tests prove otherwise an insulated protective conductor should be installed inside the trunking. It is in any case common practice now to provide separate circuit protective conductors to ensure earth continuity throughout the life of the installation. IEE Table 54G gives the sizes of protective cables required, and it will be seen that the size depends upon the size of the largest cable contained in the trunking. As with conduit the cable capacities of trunking can be calculated. To ensure that cables can be readily installed, a space factor of 45% should be used.

When a large number of cables are installed in trunking, due regard must be paid to temperature rise due to bunching of cables. IEE Table 4B1 gives details of the factors to be taken into account when cables are bunched in trunking or conduits, and in some circumstances this could result in a very considerable reduction in the current ratings of the cables installed in the trunking.

For example, if 8 circuits are enclosed in trunking the correction factor, according to IEE Table 4B1, could be as much as 0.52 to the rating values for 16 single core cables.

The ratings of cables installed in trunking are also affected by ambient temperatures, and a derating of PVC insulated cables will

Fig. 9.2 *Trunking used for fluorescent lighting in an industrial location (Thorn Lighting Ltd)*

be necessary if the ambient temperature exceeds 30°C, as will be seen by referring to the rating factors in IEE Table 4C.

Details of the application of correction factors for grouping and ambient temperature are given in Chapter 2.

Lighting trunking system

Steel or alloy lighting trunking was originally designed to span trusses or other supports in order to provide an easy and economical method of supporting luminaires in industrial premises at high levels.

The first types of such trunking consisted of extruded aluminium alloys, the sections of which were designed to support the weight of luminaires between spans of up to 5 m. More recently sheet-steel trunking has become available, made in sections which achieve the same purpose.

The advantage of this type of trunking is that it can be very easily installed across trusses, will accommodate all the wiring to feed the lighting points, and can also accommodate power wiring and, if fitted with more than one compartment, fire-alarm and extra low voltage circuits.

When installed at high levels it can be very usefully employed to accommodate wiring for high-level unit heaters, roof fans and similar equipment. Its main purpose is of course to support luminaires, and when suspended between trusses, which have a maximum spacing of 5 m, it should be able to support the weight of the required number of luminaires without intermediate supports.

It is therefore necessary that trunking suspended in this manner is of sufficient size to take the necessary weight without undue deflection. Manufacturers of trunking should be consulted about this.

This type of lighting trunking is also manufactured in lighter and smaller sections which can be fixed direct to soffits, either on the surface or mounted flush with the finished ceiling; as this does not have to support heavy weights between spans it is similar to ordinary cable trunking.

Like all other trunking, it must be provided with suitable copper links between sections to ensure adequate earth continuity, but as already explained, if the earth continuity is found to be unsatisfactory, an insulated protective conductor should be installed in the trunking.

Some types of lighting trunking are of sufficient dimensions to accommodate the fluorescent lamps and control gear within the trunking.

Non-metallic trunking

A number of versatile plastic trunking systems have been developed in recent years and these are often suitable for installation work in domestic or commercial premises, particularly where rewiring of existing buildings is required. The trunkings can be surface mounted and if care is taken in the installation, can be arranged to blend unobtrusively into the decor. Skirting-mounted trunking is probably the most appropriate for use in domestic dwellings, but shallow multi-compartment trunking can also be run at higher levels in, for example, kitchens. Industrial non-metallic trunking is also available in a range of sizes up to 150 × 150 mm. The manufacturers of plastic trunking generally supply a full range of fittings and accessories for their systems, and in some cases these are compatible between one make and another. Generally, however, once one system is chosen, it will be necessary to stay with it to achieve a neat appearance and the interchangeability of fittings.

The IEE Regulations which apply to metal trunking also generally apply to non-metallic type. Low voltage insulated or sheathed cables

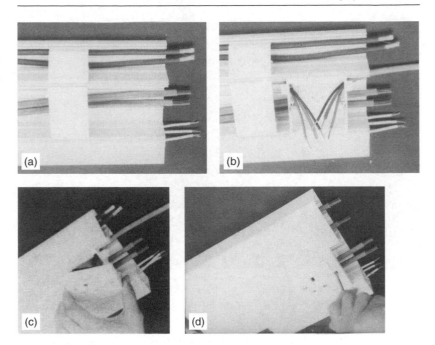

Fig. 9.3 *Stages in the assembly of a modern multi-compartment plastic trunking system. (a) Having secured the main carrier to the wall, the cables are run as required in the various compartments. Cable retainers are clipped in position as the work progresses to hold the cables in place. (b) Outlet boxes for socket are clipped in position, the appropriate cables are fitted, the cable ends stripped and earth sleeves fitted. The covers may be fitted in position at this stage, these butting up to the edges of the outlet boxes. (c, d) The sockets are connected and secured in position (MK Ltd)*

may be installed in plastic trunking. In any area where there is a risk of mechanical damage occurring, the trunking must be suitably protected. Being non-conductive, it will be necessary to run protective conductors for circuits requiring them inside the trunking, and the size of these protective conductors must be calculated so as to satisfy the IEE Wiring Regulations.

The advantages of non-metallic trunking are that it is easier to install, is corrosion resistant and is maintenance free. In addition the flexibility is such that it is often possible to reposition outlets or make other alterations without any major disturbance. The disadvantages are that there is no metal screening between low voltage, telephone, or other cables which may be run in the different compartments and

Fig. 9.4   *A typical office installation where data, telecommunication and power circuits are required. (W.T. Parker Ltd)*

there are limits to the ambient temperature in which the system can be installed.

Installation of non-metallic trunking

Care and good workmanship are needed to ensure a successful installation, and the use of good quality materials necessary. The installation layout must be planned before commencing work. If the installation is in a new or altered building, all internal structural and wall finishes should have been completed.

As with plastic conduit, it is necessary to allow for expansion of the trunking. This is done by leaving gaps between trunking sections as they are installed. A gap of 4 to 6 mm per 3 m length is recommended if high ambient temperature variations are likely to occur. The gaps are generally covered by pieces designed for the purpose. The detail will vary according to the particular system being used and the manufacturer will be able to advise on the recommended method.

The trunking should be cut using a fine tooth saw. Clean off any burrs and swarf after making the cut. Appearance will be spoiled if the cut angles do not match exactly so it is advisable to use a mitre box to make the cuts.

The main component of the trunking is generally fitted to the surface of the wall using dome-headed screws. It is essential to use washers under the screw heads, and to cater for expansion of the

Fig. 9.5   *This photographer's studio is fitted with wiring in plastic trunking feeding a busbar distribution system. The busbars incorporated in the trunking enable the socket and other power outlets to be easily fitted by plugging in. The trunking incorporates additional compartments in which other circuits can be run (MK Ltd)*

plastic components, oversize holes should be drilled in the trunking. Trunking should be fixed at intervals of not more than 500 mm, and there should also be fixings within about 100 mm of the end or of any joint. If it is intended to fit any load-bearing components such as light fittings, extra fixings should be provided. It is best to first drill the clearance holes in the trunking, and then use the prepared length as a template to mark the wall for drilling. It is possible to use shot-fired masonry pins to secure the trunking if desired. In this case it is essential to use cushioning washers under the heads of the pins.

In general, the various components of trunking systems clip together, but it may be necessary with some systems to employ glued joints. Special solvent adhesives are available for this purpose and should be applied in the same way as described in the section on installation of plastic conduit.

Once the trunking has been fixed, the cables can be run. Some makers supply special cable retaining clips which make it easier to retain cables prior to fitting the lids. Alternatively, it is a good idea to use short offcut sections of trunking lid for this purpose. Cable capacities are calculated in the same way as for conduit using a 'unit system'. The manufacturer of the trunking should be consulted for factors for other shapes.

When fitting the trunking compartment lids, increased stability and improved appearance will be achieved if the lid joints are arranged *not* to coincide with the joints in the main carrier.

Cable ducts

Cable ducting is defined in the IEE Regulations as 'a manufactured enclosure of metal or insulating material, other than conduit or cable trunking, intended for the protection of cables which are drawn in after the erection of the ducting'.

Cable ducts usually consist of earthenware or concrete pipes buried in the slab or ground, with suitable manholes to enable cables to be drawn in. IEE Regulation 522-08-03 requires that every bend formed shall be such that cables will not suffer damage. Cables installed in underground ducts should have a sheath or armour to resist any mechanical damage. Unsheathed cables must not therefore be installed in these ducts. Mineral insulated copper sheathed cables which are installed in ducts must have an overall covering of PVC sheath.

The space factor of ducts must not exceed 35%, whereas the space factor for trunking is 45%, and that for conduit is 40%. All of these space factors depend upon not more than two 90° bends (or the equivalent) being installed between draw-in points. IEE Regulation 528-01 makes it clear that Band I and Band II cables must not be installed in the same duct.

One method of forming concrete ducts is by means of a flexible rubber or plastic tubing of the required diameter. This is inflated and placed in position before the concrete slab is poured. After the concrete has set, the tube is deflated and withdrawn, and can be reused to form other ducts. Bends in ducts can be formed by this method providing the inner radius is not less than four times the diameter of the duct.

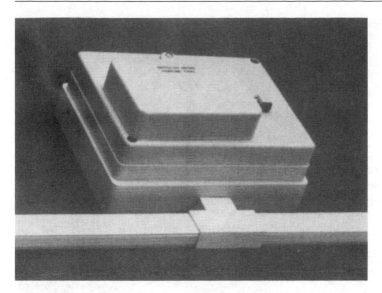

Fig. 9.6 *Mini-trunking connected to a consumer unit*

Underfloor trunking systems

Open plan office and other types of commercial buildings may well need power and data wiring to outlets at various points in the floor area. The most appropriate way of providing this is by one of the underfloor wiring systems now available. Both steel and plastic construction trunking can be obtained, and if required 'power poles' can be inserted at appropriate locations to bring the socket outlets to a convenient hand height. With the increasing use being made of computers, and other electronic data transmission systems, the flexibility of the underfloor wiring can be used to good advantage.

Steel floor trunking

Underfloor trunking made of steel is used extensively in commercial and similar buildings, and it can be obtained in very shallow sections with depth of only 22 mm, which is very useful where the thickness of the floor screed is limited.

It is supplied with one or more compartments, and with junction boxes that have cover plates fitted flush with the level of the finished floor surface. Where there are two or more compartments these boxes are fitted with flyovers to enable Band I and Band II circuits to be kept segregated as required by IEE Regulation 528-01.

Fig. 9.7   *A lighting installation in an office situation incorporating trunking to support and feed the lighting fittings. (Thorn Lighting Ltd)*

When floor ducts are covered by floor screed it is necessary to ensure that there is a sufficient thickness of screed above the top of the ducts to prevent the screed cracking as a result of the expected traffic on the floors. Another method is to use floor trunking, the top cover of which is fitted flush with the finished floor surface. In this case the top cover plate has to be of sufficient thickness to form a load bearing surface.

Outlets for sockets and other points can be fitted on top of the cover plates, and it is usual to fit pedestals to accommodate the sockets.

A more recent innovation is for the trunking to be of sufficient depth to accommodate the socket and plugs, together with the necessary wiring. The minimum depth for this type of trunking is 50 mm. Separate short sections of cover plate are provided in all positions where sockets may be required; these sections are easily

Fig. 9.8 *Power poles can be installed in office installations to bring socket outlets to convenient hand height. This pole includes both data and power circuits (Gilflex Ltd)*

removable and are provided with bushed holes to enable flexible cords to emerge. It is necessary to provide suitable holes in lino or carpets for the flexible cords to pass through.

Whatever type of floor trunking is employed, it can be connected to distribution board positions, and also to skirting trunking. Special right-angle bends are available to facilitate connection between floor trunking and skirting trunking.

If there is any doubt as to the continuity between sections of floor trunking it is advisable to run an insulated protective conductor in the trunking. Protective conductors must connect from the trunking to earthing terminals of socket outlets and other accessories.

Where socket outlets are required in positions where there is no floor trunking or skirting trunking, such points can be wired in conduit connected to the side of the trunking.

Another type of metal floor trunking is the 'In-slab' installation method. This consists of enclosed rectangular steel ducts (usually 75 mm × 35 mm), together with junction and outlet boxes.

A separate duct is provided for each wiring system, i.e. for low-voltage circuits, fire alarms, telephone lines, etc.

The separate ducts are spaced apart to give a stronger floor slab. The depth of the trunking and outlet boxes together with their supporting brackets equals that of the floor structure, so there is no need for a finishing screed, thus affording a considerable saving in construction costs.

The outlet boxes can be fixed in any position, but a distance of 1.5 m between boxes will usually provide facilities for most office needs.

Fig. 9.9   *Plastic trunking is available with transparent lids. This is useful in situations where security checks on circuits need to be made, and provides for the rapid identification of circuits and circuit integrity (Gilflex Ltd)*

Plastic underfloor trunking

As with many other types of wiring system available such as conduit or trunking, plastic materials are now often used instead of their metal counterparts for the enclosures of underfloor systems.

Underfloor trunking systems made with this material can be divided into two main types, these being raised floor systems and underfloor ducted systems.

The raised floor installation has the advantage of extreme flexibility as the load bearing floor is structurally supported such that there is an unobstructed space underneath. The wiring ducts can thus be run under the floor in any desired position. The outlet positions which are incorporated in floor panel sections are connected to the

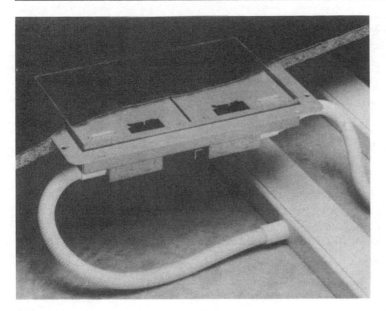

Fig. 9.10 *A cut-away view of underfloor trunking installed in a raised floor system. Flexible conduit will permit limited repositioning of the outlets when needed (Gilflex Ltd)*

Fig. 9.11 *Underfloor non-metallic trunking installed in a commercial office installation. With the growth of data processing, flexible office wiring systems are a necessity, and a raised floor provides a viable method of achieving this (MK Ltd)*

Fig. 9.12 Underfloor connections between trunking and the floorboxes. Power and data cabling can be seen, and the floor outlet boxes can be repositioned if office layout changes occur (MK Ltd)

Fig. 9.13 A floor outlet box with segregated sections for power and telecommunications wiring (MK Ltd)

ducted wiring using flexible conduit and in this way outlet positions can be rearranged at will by exchanging the floor panel sections. This type of layout is especially useful in computer rooms where due to the rapid advance of technology it is necessary to replace obsolete equipment at intervals.

The other system supplied in plastic materials is the underfloor ducted system. With this, shallow ducts are installed prior to the final floor surface being laid. The ducting is subsequently buried in the concrete screed. A variety of outlet positions can be used. Concealed and raised socket outlets are available, and as previously mentioned, 'power poles' can also be fitted. Some manufacturers supply fittings whereby connection can readily be made to skirting trunking.

10   *Busbar and grid catenary systems*

Busbar system

The busbar system can be used for 3-phase distribution to machines in factories. This consists of copper or aluminium busbars mounted on insulators and enclosed in standard lengths of steel trunking, which are arranged to be bolted together thus forming a continuous busbar along the entire length of factory or shop.

At intervals, usually every 1 m, a tap-off point is provided. At these points tap-off units may be fitted; these consist of three HRC fuses mounted in a sheet case with a hinged door (Figs. 10.1 and 10.2).

The units are provided with contact fingers which are designed to fit on to the busbars.

Connections from these tap-off units to motors, welders or other electrical equipment can be made by flexible connections, PVC sheathed cables, or conduit.

The advantages of this system are that the trunking and busbars can be erected before the installation of the machinery, and the latter can be connected up and set to work as soon as they are installed.

By bringing the heavy main feeders near to the actual loads, the circuit wiring is reduced to a minimum and voltage drop is lower than would otherwise be the case.

Subsequent additions and alterations to plant layout can be easily accomplished, and where busbar sections have to be removed they can be used again in other positions.

The overhead busbar system is especially advantageous where a large number of electric welders have to be fed with heavy currents from a step-down transformer.

If a large number of small machines are to be fed it is usual to fit a distribution board near the trunking system and to protect this with a tap-off box fitted with HRC fuses of suitable capacity. Circuit wiring from the distribution board is usually carried out in heavy gauge screwed conduit.

Fig. 10.1   *A plug-in connection unit for use with a busbar trunking system. These plug-in units can be supplied with isolating switches and/or fuses (BP ltd)*

Fig. 10.2   *Overhead busbar system. The plug-in fuseboxes are designed for making connection with live busbars and can be inserted at a number of points in the run (GEC Installation Equipment Ltd)*

The system is comparatively expensive in first cost and, therefore, is best employed where heavy loads and a large number of machines have to be provided for. Once installed there is very little depreciation or need for maintenance, and it has a high recovery value should it be necessary to dismantle and install elsewhere.

Several well-known manufacturers supply this trunking in various sizes, together with all necessary tees, bends, tap-off boxes and other accessories. Earth continuity is usually provided by an external copper earth link which ensures good continuity.

Where there are long runs of busbar trunking it is necessary to provide expansion joints to take up any variations in length due to changes in temperature. These expansion joints usually take the form of a short length of trunking enclosing flexible copper braided conductors instead of solid busbars. It is advisable to provide one of these in every 30 m run of trunking.

No other conductors of any kind may be installed inside trunking containing bare copper busbars.

All lids and covers must always be kept in position as a protection against vermin and also to avoid accidental contact with live busbars. The trunking should be marked prominently on the outside at intervals with details of the voltage of the busbars and the word DANGER.

The conductors shall be installed so that they are not accessible to unauthorised persons.

Insulators shall be spaced to prevent conductors coming in contact with each other, with earthed metal or other objects.

The conductors shall be free to expand or contract during changes of temperature without detriment to themselves or other parts of the installation. In damp situations the supports and fixings shall be of non-rusting material.

Conductors shall not be installed where exposed to flammable or explosive dust, vapour or gas, or where explosive materials are handled or stored.

Where the trunking passes through walls or floors no space shall be left round the conductors where fire might spread. Fire barriers should be provided at these points inside the trunking.

All runs of overhead busbar trunking must be capable of isolation in case of emergency or maintenance by means of an isolating switch fixed in a readily accessible position (Electricity at Work Regulations 1989 – Regulation 12).

Busbar system for rising mains

A similar busbar system is frequently used for vertical rising mains for multistorey buildings. This usually consists of copper or aluminium busbars of capacities of 100 A to 500 A with two, three or four conductors. These are usually metalclad and are made in 4 m sections, although all-insulated rising busbar systems are also obtainable.

Tap-off boxes with fuselinks or fuseswitches can be provided for distribution to each floor where distribution boards can be fitted near the tap-off units (Fig. 10.3). For these vertical runs it is very important that fire resisting barriers be fitted inside the trunking at the level of each floor. These fire barriers can be purchased with the trunking, and the manufacturers will fit these in the required positions if provided with the necessary details. Where the trunking passes through floors, whether in a specially formed riser cupboard, or run on the surface of a wall, it is necessary to ensure that the floor is 'made good'

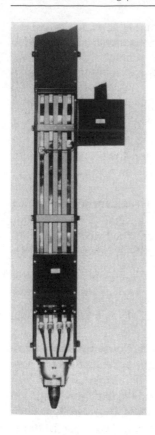

Fig. 10.3　*300 A, 4-pole rising busbars with 100 A s.p.n. tap-off unit (Davis Trunking Ltd)*

by non-combustible material round the outside of the trunking to prevent the spread of fire.

The supply for these runs of busbars is usually effected by a feeder box with provision for whatever type of cable is used. The manufacturers should be consulted as to the correct size of busbars to use, and IEE Regulations recommend that the maximum operating temperature should not exceed 90°C. Where rubber or PVC cables are connected to busbars operating at comparatively high temperatures the insulation and sheath must be removed for a distance of 150 mm from the connection and replaced by suitable heat resisting insulation.

Grid catenary system

The catenary system of wiring is designed for large buildings or open areas where a screwed conduit system would be difficult and expensive to install.

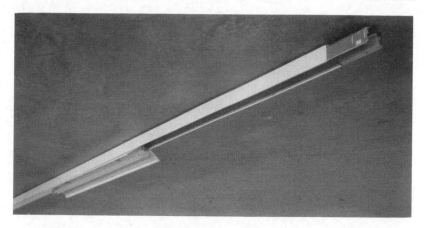

Fig. 10.4 *This type of lighting trunking has integral busbars carrying conductors for the supply to the luminaires. The 54 watt twin fittings are available with either wide angle or semi-specular narrow beam characteristics, enabling effective lighting from different heights. The trunking can be suspended from the building frame and, with the narrow beam reflectors, is suitable for high-bay illumination. The clip-in fittings can be arranged to pick up from different phases and the whole assembly can be carried out without the use of tools, a particular advantage when working at heights. (W.T. Parker Ltd)*

This cable consists of a number of PVC insulated cores made up round a high-tensile galvanised steel wire, with suitable fillings of PVC to produce a circular construction; the whole is sheathed overall in PVC and so rendered waterproof.

This cable may be suspended between roof members by means of the steel catenary wire secured by eye bolt and turnbuckles.

One standard connection box is used (Fig. 10.5), and this box may be used for tee, angle or through joints. These same boxes may also be used as boxes from which luminaires can be suspended.

All kinds of special accessories are available to enable these boxes and turnbuckles to be fitted to trusses and other structural steelwork and brick walls.

The cables can be suspended in spans up to 60 m, but it is usual to provide auxiliary suspension devices at more frequent intervals, generally at trusses and also immediately above lighting luminaires.

A complete section can be made up at ground level, where connecting boxes, luminaires, lamps, etc., can be assembled complete before erection; this saves time and avoids the difficulty of working in high positions.

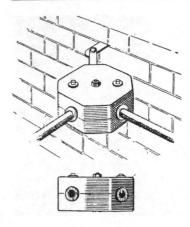

Fig. 10.5   *Typical corner assembly. Box has been adapted for compound filling by fitting plugs*

Standard cables are supplied in 4-core from 1.5 mm^2 to 16 mm^2 sizes, and the standard connection boxes are made to accommodate these standard sizes. 5- and 7-core cables can also be supplied in certain sizes.

The connecting boxes also serve as anchor boxes and are designed to anchor the steel cable at terminations, at right angles and also at mid-span positions. All metal catenary wires must be earthed. In outside installations, or where exposed to steam or other adverse conditions, the connecting boxes must be filled with a plastic compound to render them weatherproof. Where the system is to be installed in atmospheres aggressive to aluminium alloy, the grid suspension boxes can be supplied with a protective PVC coating.

11 Power cable systems

At one time, paper insulated lead-covered cables were almost exclusively used for main and submain cables. For cables buried in the ground, lead alloy provides excellent corrosion resistance. However, a range of other cable types is now available, and these include PVC insulated, XLPE insulated and LSF type, usually with wire armouring. Each has characteristics which can be appropriate to a range of installation situations, and some detail is given in the sections which follow.

Armoured, insulated and sheathed cables

Armoured PVC and XLPE insulated cables are now being used extensively for main cables and distribution circuits, and also for circuit wiring in industrial installations.

These cables consist of multicore PVC insulated cables, with PVC sheath and steel wire armour (SWA), and PVC sheathed overall. Their advantage over PILC cables is that they are more pliable, and can be bent to a radius of only eight times their diameter, against twelve times required for PILC and armoured cables. They are much lighter, and easier to handle, and the sealing of the ends is much simpler.

The main disadvantages are that thermo-plastic insulation will sustain serious damage if subject to temperatures over 70°C for a prolonged period, and proper protection against sustained overloads is required. The insulation will harden, and become brittle in temperatures below 1°C, and the cables should not be installed or handled when temperatures are approaching freezing, otherwise the insulation may be inclined to split. Low temperatures will do no permanent harm to the insulation, providing the cables are not interfered with during extreme cold.

PVC/SWA/PVC multicore sheathed cables are manufactured in all sizes up to 400 mm^2.

Details of current ratings are given in Table 3.3. These cables can be laid directly in the ground, in ducts, or fixed to the surface on a cable tray, or fixed to the structure by cleats. When a number of multicore cables take the same route, it is an advantage for them to be supported on cable trays or ladders, which are manufactured in various sizes from 100 mm to 1 m wide.

The current-carrying capacity of PVC/SWA/PVC cables, like that of PILC cables, is to some extent determined by the method of installation: for example, by reference to Table 3.3 it will be seen that cables installed under method 11 have a higher rating than for cables installed under method 1.

When several cables are grouped together on a wall, or tray, or in ducts, the current rating will have to be reduced according to the correction factors as described in Chapter 2.

The smaller multicore cables have many advantages when used in industrial installations for circuit and control wiring owing to their ease of installation, flexibility, and high recovery value when alterations become necessary.

End terminations are made by stripping back the PVC sheathing, and steel wire armouring, and fitting a compression gland which can be screwed to switchgear, etc., and provides earth continuity between the armour of the cable and the switchgear. When connecting to motors on slide rails, a loop should be left in the cable near the motor to permit the necessary movement.

XLPE cables

XLPE cables are made to BS 5467 and are becoming increasingly used in new installations. XLPE has better insulation qualities than PVC and thus it is possible to obtain cables of a smaller diameter for the same voltage rating. In addition, provided suitable terminations are used, XLPE cable may be used at a maximum working temperature of 90°C. These cables are available in sizes up to 400 mm^2 or 1000 mm^2 single core.

Jointing must be given careful consideration, and crimping is recommended rather than soldering if advantage is to be taken of the maximum short circuit capability of the cable. Also, the jointing compound used must be selected to suit XLPE, and some materials such as PVC tape are incompatible and must not be used. Compounds which suit LSF as well as XLPE are available.

LSF sheathed cables

In situations where there is a need to protect people who may be at risk due to the outbreak of fire, low smoke and fume (LSF) cables to

Fig. 11.1 *A typical section of heavy-duty cable tray (Airedale Sheet Metal Ltd)*

BS 6724 may be used. In certain circumstances, emissions of toxic gases are reduced, and the cables are slow to ignite. This reduces the risk to occupants, and increases the ability to escape.

Locations where this may be relevant include underground passageways or tunnels, cinemas, hospitals, office blocks and other similar places where large numbers of people may be present. Cables are available in sizes up to $630\,mm^2$ single core, and $400\,mm^2$ two, three or four core. Typical cables have copper conductors, XLPE insulation with LSF bedding, single wire armouring and an LSF sheath. Control cables with LSF insulation can also be obtained in sizes up to $4\,mm^2$ and with up to 37 cores.

Cable tray, cable basket and cable ladder

Cable tray, cable basket and cable ladder are ideal methods of supporting cables in a variety of situations. With care, a very neat appearance can be obtained, and with both vertical and horizontal runs, cables can be run in line, free from deviations round beams or other obstructions. In buildings with suspended ceilings, cable tray offers an ideal method of running wiring in the ceiling void.

Cable tray comprises the basic lengths of galvanised steel tray, usually in 3 m sections, and a range of fixings to enable the sections to be joined and run round vertical or horizontal bends. Various support accessories are also available, and the accompanying illustrations show the steps needed in the assembly of a typical run. Cable

Fig. 11.2 *Cable runs fitted in the roof space above a suspended ceiling. Multi-compartment cable basket is provided, as well as cable tray for different circuit categories. Note the copex flexible conduit which carries feeds to specific locations (W.T. Parker Ltd)*

Fig. 11.3 *A cable tray bending machine which is capable of handling cable tray up to 24 inches in width (Hilmor Ltd)*

tray is cut using a hacksaw and after cutting the burrs are removed with the use of a file. The sections are joined using joining strips which are bolted in place with galvanised bolts and nuts. Some makes of tray may be joined by the use of spring clips. Light duty tray may be bent by hand after cutting the side flange, and special bending machines deal with bending of heavy duty cable tray (Fig. 11.3).

Light and heavy section cable tray is available, and for some of the heavier duty types, there can be a distance of several metres between

Fig. 11.4 *Installation of cable tray. Supports appropriate to the installation are first fitted. Here ceiling supports to carry cable tray and trunking above a suspended ceiling can be seen*

Fig. 11.5 *After measurement and marking off, the length of cable tray is cut using a hacksaw*

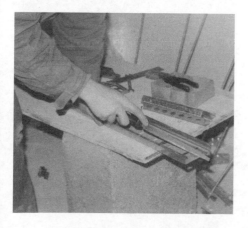

Fig. 11.6 *Using a file, burrs are removed to prevent damage to the cables*

Fig. 11.7   *Joining strips are installed by sliding them over the tray flange, and bolting into position. This is carried out at one end of each length before erection. Then the cable tray is lifted into position and the flanges slid onto the adjoining length, and in turn bolted*

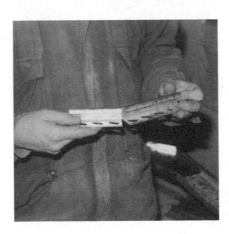

Fig. 11.8   *Where a 45° bend is required a similar method of joining is used, but the joining strip must be bent after cutting through the flange*

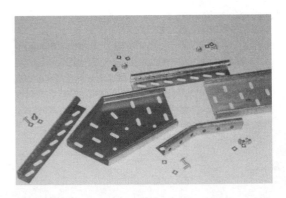

Fig. 11.9   *The components of a 45° bend ready for assembly*

Fig. 11.10 *As with straight joints, the joining strips are bolted in position. With cable tray which is run in inaccessible positions, as much work as possible should be done at the bench, leaving only the final bolting of sections to be done on site*

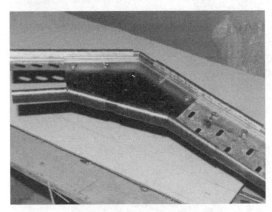

Fig. 11.11 *The finished bend ready for erection*

Fig. 11.12 *For vertical bends or folds, both flanges of the cable tray must be cut at a suitable angle. This view shows one complete cut, the second flange having not yet been dealt with*

supports. The use of such trays enables considerable savings to be made in fixing costs.

Sheathed and/or armoured cables which are run on cable trays need not be fixed to the tray providing the cables are in inaccessible positions and are not likely to be disturbed, and that the cables are neatly arranged in such a manner that the route of each cable can be easily traced.

In many industrial buildings the roof purlins are specially shaped to accommodate multicore and other types of cable, thus eliminating

Fig. 11.13 *Finally, the cable tray is lined up neatly in the support brackets, and bolted into position (William Steward & Co. Ltd)*

Fig. 11.14 *A neat installation of cable tray in a substation carrying feeds to individual distribution boards (W.T. Parker Ltd)*

Fig. 11.15 *Submain distribution cables run on cable ladder and cleated. The PVC/SWA/PVC cables shown are 4-core 240 mm² and feed four switchboards in a hospital. Each switchboard has separate feeds for essential and non-essential supplies (hence eight cables), and in the event of mains failure, the essential circuits are fed by diesel driven emergency alternators (William Steward & Co. Ltd)*

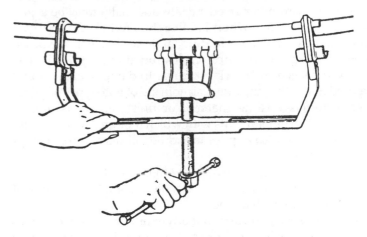

Fig. 11.16 *Combined bending and straightening tool for aluminium-sheathed cables exceeding 30 mm diameter*

the need for any additional method of support or fixing. The 'Multi-beam' system comprises purlins specially designed to accommodate cables, and various accessories are available to provide for outlets for lighting fittings and power points. Where unsheathed PVC cables are installed in these purlins, insulated covers are provided to give the necessary protection against mechanical damage.

Paper-insulated cables

Paper-insulated lead covered cables are seldom used today but may occasionally be encountered in old installations.

This type of cable may be encountered in use in distribution systems in factories and between factory buildings, and when so used they come within the scope of IEE Regulations.

The terminations and joints of PILC cables should be protected from the ingress of moisture by being suitably sealed. Due to the special skills required for jointing these cables they are being rapidly replaced by PVC insulated and armoured cables.

The maximum internal radii of bends in PILC cables shall not be less than twelve times the overall diameter of the cable.

Single-core cables armoured with steel wire or tape shall not be used for a.c. (IEE Regulation 521-02), but single-core cables with aluminium sheaths may be used providing the current ratings in IEE Table 4K1 are complied with, and that suitable mechanical protection is provided where necessary. Single-core cables are not usually armoured.

Metal armouring of PILC and PVC/SWA cables which come into fortuitous contact with other fixed metalwork shall either be segregated therefrom or effectively bonded thereto. Both PILC and PVC/SWA armoured cables shall have additional protection where exposed to mechanical damage; for example cables run at low levels in a factory might be damaged by a fork lift truck. In damp situations, and where exposed to weather, the metal armouring of both types of cables shall be of corrosion-resisting material or finish, and must not be placed in contact with other metals with which they are liable to set up electrolytic action. This also applies to saddles, cleats and fixing clips (IEE Regulation 522-05).

For internal situations it is permissible to use lead sheathed PILC cables without further protection if installed at high levels out of reach and not likely to be disturbed.

If it becomes necessary to carry out any work on multicore cables that have already been in service, precautions must be taken to ensure that no current is present in the cable due to its capacitance. Long runs of cable act as capacitors when in service, and when disconnected from the source of supply a high potential may have been built up in the cable. Before touching any of the conductors, therefore, the current should be discharged by connecting a 100 W lamp between earth and each conductor in turn.

In the case of underground multicore cables that have been in service and have to be cut, it is usual to spike the cable with a metal spike at the position where it is proposed to cut it. The spike should penetrate the earthed sheath and all the conductors; this will ensure that the cable is discharged, and will also obviate any risk of accidentally cutting through a cable that may be live.

12 All-insulated wiring systems

The first all-insulated wiring system consisted of vulcanised insulated conductors sheathed with a tough rubber sheath (TRS). When first introduced the system was known as the 'cab-tyre' system (CTS), because the outer sheath resembled the type of hard rubber used in the manufacture of solid tyres for horse-drawn cabs. The TRS (or CTS) system has now become almost obsolete, as it has been replaced by the PVC insulated and sheathed system.

The PVC system has many advantages over the old TRS system because it is not so flammable, and will stand up better to direct sunlight and chemical action. The cables may be installed without further protection, except where exposed to mechanical damage, when they must be suitably protected.

This all-insulated wiring system is used extensively for lighting and socket installations in small dwellings, and is probably the most economical method of wiring for this type of work. It is customary to use 2- and 3-core cables with an integral protective conductor and to provide insulated joint boxes or 4-terminal ceiling roses for making the necessary connections (Figs. 12.1 to 12.4).

An alternative method of wiring with PVC sheathed cables for lighting is to use 2-core and cpc cables with 3-plate ceiling roses instead of joint boxes (Fig. 12.5).

IEE Regulation 526-03-02 requires that terminations or joints in these cables must be enclosed in non-ignitable material, such as a box complying with BS 476 part 12, or an accessory or luminaire. (An 'accessory' is defined as 'a device, other than current-using equipment, associated with such equipment or with the wiring of an installation'.)

At the positions of joint boxes, switches, sockets and luminaires the sheathing must terminate inside the box or enclosure, or could be partly enclosed by the building structure if constructed of incombustible material.

Fig. 12.1   *Insulated joint box for PVC-sheathed wiring system, showing three 2-core and cpc cables connected to box. All joints must be made in a box or at an accessory or luminaire. With solid conductor cables special care must be taken to ensure that all conductors are firmly gripped. It may be necessary to double back one conductor to obtain a satisfactory disposition of the conductors in the terminal tunnel. The protective conductor must be fitted with a green/yellow sleeve*

Fig. 12.2   *Ceiling rose with looping and earth terminals (MK Ltd)*

Surface wiring

When cables are run on the surface a box is not necessary at outlet positions, providing the outer sheathing is brought into the accessory or

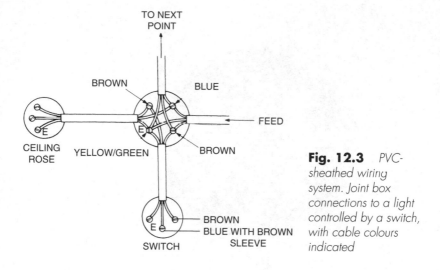

TO NEXT
POINT

BROWN   BLUE

FEED

CEILING
ROSE   YELLOW/GREEN   BROWN

E

BROWN
BLUE WITH BROWN
SLEEVE

SWITCH

Fig. 12.3   *PVC-sheathed wiring system. Joint box connections to a light controlled by a switch, with cable colours indicated*

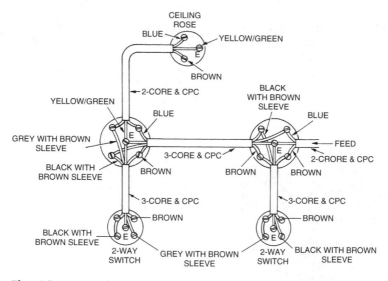

CEILING
ROSE

BLUE   YELLOW/GREEN

E

BROWN

2-CORE & CPC

YELLOW/GREEN   BLACK
WITH BROWN
SLEEVE

BLUE   BLUE

GREY WITH BROWN
SLEEVE   FEED

3-CORE & CPC   2-CRORE & CPC

BLACK WITH
BROWN SLEEVE   BROWN   BROWN   BROWN

3-CORE & CPC   3-CORE & CPC

BROWN   BROWN

BLACK WITH
BROWN SLEEVE
E

2-WAY
SWITCH   GREY WITH BROWN
SLEEVE   2-WAY
SWITCH   BLACK WITH BROWN
SLEEVE

Fig. 12.4   *Joint box connections to two 2-way switches controlling one light*

luminaire, or into a block or recess lined with incombustible materials, or into a plastic patress.

For vertical-run cables which are installed in inaccessible positions and unlikely to be disturbed, support shall be provided at the top of

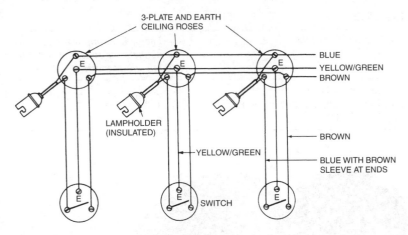

Fig. 12.5 *Method of wiring with PVC-sheathed cables using 3-plate and earth ceiling roses instead of joint boxes*

Table 12.1 *Minimum spacing of supports for non-armoured PVC insulated and sheathed cables. These spacings are for cables in accessible positions*

Overall diameter of cable (mm)	Generally		In caravans	
	Horizontal (mm)	Vertical (mm)	Horizontal (mm)	Vertical (mm)
Not exceeding 9	250	400	250	400
Exceeding 9 and not exceeding 15	300	400	250	400
Exceeding 15 and not exceeding 20	350	450	250	400
Exceeding 20 and not exceeding 40	400	550	250	400

the cable, and then at intervals of not less than 5 m. For horizontal runs the cables may rest without fixings in positions which are inaccessible and are not likely to be disturbed, provided that the surface is dry, reasonably smooth and free from sharp edges.

In caravans the fixing spacing for cables installed in accessible positions is 250 mm for horizontal runs and 400 mm for vertical runs (see IEE Regulation 608-06-05) and providing the walls or ceilings are constructed of non-combustible materials.

The minimum radii of bends in PVC cables are given in Table 12.2.

Table 12.2

Overall diameter of cable	Factor to be applied to overall diameter of cable to determine minimum radius
Not exceeding 10 mm	3
Exceeding 10 mm, but not exceeding 25 mm	4
Exceeding 25 mm	6

Fig. 12.6   *13 A ring circuit joint box. For flat PVC cables up to 6 mm², with selective shutters on cover closing the unused entries (Ashley Accessories Ltd)*

PVC and similar sheathed cables if exposed to direct sunlight shall be of a type resistant to damage by ultraviolet light (IEE Regulation 522-11). PVC cables shall not be exposed to contact with oil, creosote and similar hydrocarbons, or must be of a type to withstand such exposure (IEE Regulation 522-05).

Concealed wiring

PVC wiring, concealed in floors or partitions, is an effective method of providing a satisfactory installation where appearance is of prime importance as in domestic, display or office situations. Such wiring arrangements are covered by IEE Regulation 522-06-06. There is no reason why PVC sheathed cables shall not be buried direct in cement or plaster, provided the location is such that IEE Regulation 522-06-06 is complied with. The relevant locations are illustrated in Fig. 12.9. A disadvantage is that cables once buried in cement or plaster cannot be withdrawn should any defect occur, and the circuits would then have to be rewired. It is better to provide a plastic conduit to the switch or

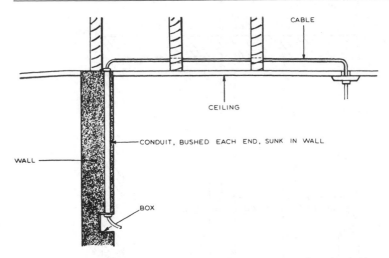

Fig. 12.7 *Concealed PVC-sheathed wiring system. When the cable is concealed in a wall a box must be provided at the switch position whether the switch is flush or surface type*

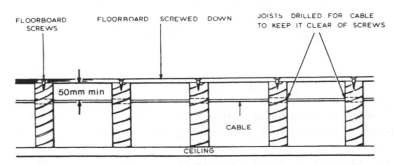

Fig. 12.8 *Running PVC-sheathed cable under wooden floors across joists*

outlet positions so that the PVC cables can be drawn into the conduit, and withdrawn should the need arise. Such an arrangement must also comply with the location constraints given in Fig. 12.9.

If it is impractical to run concealed wiring at the locations specified, then appropriate protection must be provided. This may take the form of a cable incorporating an earthed metal sheath, or by enclosing the cables in earthed metallic conduit, trunking or ducting.

Whichever construction is employed, it is necessary to provide a box at all light, switch and socket outlet positions (Fig. 12.7). The boxes must be provided with earthing terminals to which the protective conductor in the cable must be connected. If the protective conductor is a

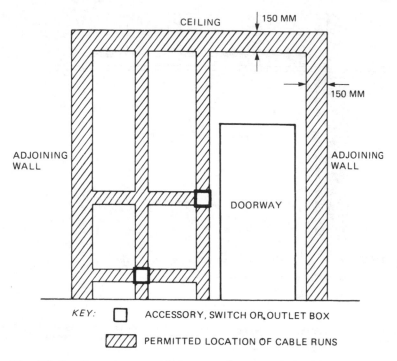

KEY: ☐   ACCESSORY, SWITCH OR OUTLET BOX

        ▨   PERMITTED LOCATION OF CABLE RUNS

Fig. 12.9   *Typical permissible locations for concealed cable runs (IEE Regulation 522-06-06). Where it is impractical to use these locations, special precautions are necessary, see text*

bare wire in a multicore cable, a green/yellow sheath must be applied where the cable enters the box (IEE Regulation 514-03-02).

Keep cables away from pipework
Insulated cables must not be allowed to come into direct contact with gas pipes or non-earthed metal work, and very special care must be exercised to ensure they are kept away from hot water pipes.

Precautions where cables pass through walls, ceilings, etc.
Where the cables pass through walls, floors, ceilings, partitions, etc., the holes shall be made good with incombustible material to prevent the spread of fire. It is advisable to provide a short length of pipe or sleeving suitably bushed at these positions, and the space left inside the sleeve should be plugged with incombustible material. Where the cables pass through holes in structural steelwork, the holes must be bushed so as to prevent abrasion of the cable.

Where run under wood floors, the cables should be fixed to the side of the joists, and if across joists, should be threaded through holes drilled through the joists in such a position as to avoid floorboard nails and screws.

Where cables are sunk into floor joists the floorboards should be fixed with removable screws.

In any case, screwed 'traps' should be left over all joint boxes and other positions where access may be necessary (Fig. 12.8).

Fixing cables by suspension on catenary wires

Another method of fixing PVC sheathed cables is by suspending them on catenary wires. Cables can sometimes be taken across a lofty building, or outside between buildings if protected against direct sunlight, by this method.

Galvanised steel wires should be strained tight and the cables clipped to the wire with wiring clips. Alternatively, they can be suspended from the wire with rawhide hangers; this provides better insulation although not so neat as the former method.

The catenary wire must be bonded to earth.

Wiring to socket outlets

When PVC cable is used for wiring to socket outlets or other outlets demanding an earth connection it is usual to provide 2-core and cpc cables. These consist of two insulated conductors and one uninsulated conductor, the whole being enclosed in the PVC sheathing. Check that the protective conduction complies with IEE Section 543.

When wiring to 13-A standard domestic sockets, the cables will have to be taken into the standard box which is designed for these sockets and which includes an earth terminal.

Practical hints

Care must be taken when stripping the sheathing of PVC cables so as to avoid nicking the inside insulation.

The sheathing must only be removed as far as necessary to enable the insulated conductors to be manipulated and connected. The sheathing must be taken well into junction boxes, switch boxes, etc., as the insulation must be protected over its entire length.

Multicore cables have cores of distinctive colours; the brown should be connected to phase terminals, the blue to neutral or common return, and the protective conductor to the earth terminal. Clips are much neater than saddles, but when more than two cables are

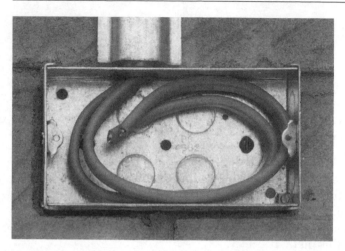

Fig. 12.10   Installation of a ring circuit in a domestic situation. The blockwork is recessed to accept the box, and the cable is installed prior to plastering. After plastering, the outer cable sheath is stripped and the socket outlet fitted (Crabtree Electrical Industries Ltd)

run together it is generally best to use large saddles. If a number of cables have to be run together on concrete or otherwise where the fixings are difficult to obtain, it is advisable to fix a wood batten and then to clip or saddle the cables to the batten.

Cable runs should be planned so as to avoid cables having to cross one another, and additional saddles should be provided where they change direction.

PVC sheathed cables should not be used for any systems where the normal voltage exceeds 1000 V.

13 Installation of mineral insulated cables

Mineral insulated (MI) cables have been in use for a sufficient number of years to have stood the test of time. These cables have an insulation of highly compressed magnesium oxide powder (MgO) between cores and sheath. During manufacture the sheath is drawn down to the required diameter; consequently the larger sizes of cable yield shorter lengths than the smaller sizes. Generally MI cable needs no additional protection as copper is corrosion resistant. However, in certain hostile environments, or if a covering is required for aesthetic or identification purposes, MI cable is available with PVC or LSF covering.

The advantages of MI cables are that they are self-contained and require no further protection, except against the possibility of exceptional mechanical damage; they will withstand very high temperatures, and even fire; they are impervious to water, oil and cutting fluids, and are immune from condensation. Being inorganic they are non-ageing, and if properly installed should last almost indefinitely.

The overall diameter of the cable is small in relation to its current-carrying capacity, the smaller cables are easily bent and the sheath serves as an excellent protective conductor. Current-carrying capacities of MI cables and voltage drops are given in IEE Tables 4J1A, 4J1B, 4J2A and 4J2B.

Fixing
The cable can be fixed to walls and ceilings in the same manner as PVC insulated cables. A minimum bending radius, of six times the bare cable outside diameter is normally applicable. This permits further straightening and re-bending when required. If more severe bends are unavoidable, they should be limited to a minimum bending radius of three times the bare cable diameter, and any further

Table 13.1   *Minimum spacing of fixings for MI copper-sheathed cables. These spacings are for cables in accessible positions*

Overall diameter of cable (mm)	Mineral-insulated copper sheathed cables	
	Horizontal (mm)	Vertical (mm)
not exceeding 9	600	1800
exceeding 9 and not exceeding 15	900	1200
exceeding 15 and not exceeding 20	1500	2000

Fig. 13.1   *Straightening up multiple runs of MI cable using block of wood and hammer*

straightening and bending must be done with care to avoid damaging the cable.

All normal bending may be carried out without the use of tools, however, two sizes of bending levers are available for use with the larger diameter cables or when multiple bends are required. These levers are specially designed to prevent cable damage during bending (Fig. 13.3).

When carrying out the installation of this wiring system the sheathing of the cable must be prevented from coming into contact with wires, cables or sheathing or any extra-low voltage system (not exceeding 50 V a.c. or 100 V d.c.), unless the extra-low voltage wiring system is carried out to the same requirements as for a low-voltage system (1000 V a.c.). This means that it must not be allowed to come into contact with lightly insulated bell or telephone cables.

Fig. 13.2   *Control wiring in 3 × 1.5 mm² PVC-sheathed cable*

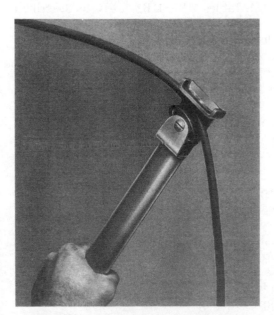

Fig. 13.3   *Bending and setting of MI cable. These operations can be more easily done by means of the simple tool illustrated (BICC Pyrotenax)*

Protection against mechanical damage

Mineral insulated cable will withstand crushing or hammering without damage to the conductors or insulation. However, if the outer sheathing should become punctured, the insulation will begin to 'breathe' and a low insulation resistance will result. Therefore it is advisable to protect the cable if there is a possibility of its being mechanically damaged.

Where cables are exposed to possible mechanical damage it is advisable to thread the cables through steel conduits, especially near floor levels, or to fit steel sheathing over the cables in vulnerable positions. Where cables pass through floors, ceilings and walls the holes around the cables must be made good with cement or other non-combustible material to prevent the spread of fire, and where threaded through holes in structural steelwork the holes must be bushed to prevent abrasion of the sheathing.

Bonding

Because of the compression-ring type connection between the gland and the cable, and the brass thread of the gland, no additional bonding between the sheath of the cable and connecting boxes is necessary although it is now common practice for the terminating pot to have an earth tail fitted. The earth continuity resistance between the main earthing point and any other position in the completed installation must comply with IEE Tables 41B1, 41B2 or 41C as described in Chapter 2.

A range of glands and locknuts is available for entering the cables into any standard boxes or casings designed to take steel conduit. The glands, which are slipped on to the cables before the cable ends are sealed, firmly anchor the cables and provide an efficient earth bonding system.

In some instances, it may not be possible to ensure bonding via the gland, e.g. when fixed into a plastic box. In these instances, a seal is available which incorporates an additional earth tail wire.

Regulations on sealing

The ends of MI metal-sheathed cables must be sealed to prevent the entry of moisture and to separate and insulate the conductors.

The sealing materials shall have adequate insulating and moisture-proof properties, and shall retain these properties throughout the range of temperatures to which the cable is subjected in service. The manufacturers provide a plastic compound for use on the standard cold screw-on pot type seal (Fig. 13.4).

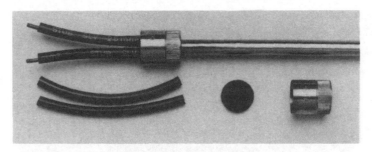

Fig. 13.4 *Mineral insulated cable screw-on pot seal (BICC Pyrotenax)*

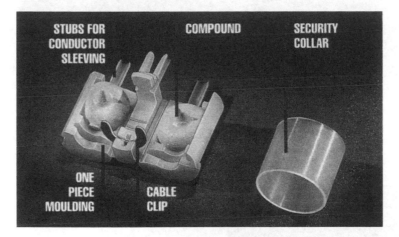

Fig. 13.5 *New seals are becoming available which incorporate the sealing compound in the moulding. Operating temperatures can be up to 80°C (BICC Pyrotenax)*

Methods of stripping and sealing are given below. The tools required include hacksaw, side cutting pliers, screwdriver, special ringing tools and pot wrench.

Preparation of cable end

To prepare the ends of the cable prior to sealing, cut the cable to length with a hacksaw (Fig. 13.5), ring the sheath with a ringing tool (Fig. 13.6), tighten the wingnut of the ringing tool so that the wheels JUST grip the sheath and then give the wingnut approximately a quarter turn (for larger cables up to half a turn). Then rotate the tool through 360° or more.

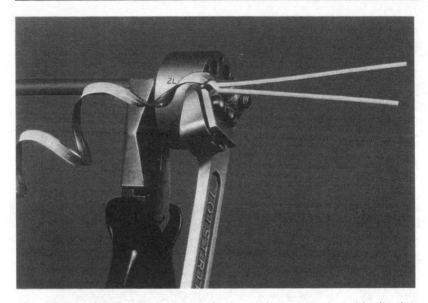

Fig. 13.6   The MI Cable Joistripper being used to remove the cable sheath (BICC Pyrotenax)

Fig. 13.7   Preparing cable end. To strip the sheath from the cable end, first use side-cutting nipper to start the 'rip'. Then proceed in a series of short rips

If the ring is made too deep it will be found difficult to break into it when stripping; if too shallow the sheath will be bell-mouthed and the gland and seal parts will not readily fit on to the sheath. If there is any roughness left around the end of the sheath from the ringing tool, remove it by lightly running the pipe grip part of a pair of pliers over it. If the length to be stripped is very long, defer ringing until stripping is within 50 to 75 mm of the sealing point.

Stripping the sheath   After ringing at the sealing point, strip the sheath to expose the conductors. Use side cutting pliers to start

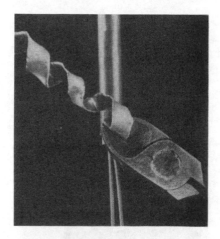

Fig. 13.8 *Preparing cable end. Strip cable end by gripping the edge of the sheath between the jaws of side-cutting nippers and twist the cable off in stages, keeping the nippers at about the angle shown*

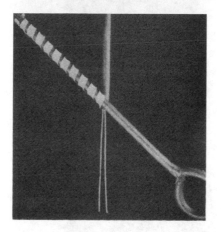

Fig. 13.9 *An alternative method of stripping sheath to expose long conductors. A stripping rod, which can be easily made from a piece of mild steel is used in a similar manner to a tin opener*

the 'rip'. To do this, grip the edge of the sheath between the jaws of the pliers and twist the wrist clockwise, then take a new grip and rotate through a small angle (Figs. 13.7 and 13.8). Continue this motion in a series of short 'rips' keeping the nippers at about 45° to the line of the cable, removing the sheath spirally. When about to break into the ring, bring the nippers to right angles with the cable. Finish off with point of nippers held parallel to the cable.

An alternative method of stripping, often employed for long tails, is to use an easily constructed stripping rod as illustrated in Fig. 13.9. This can easily be made from a piece of mild steel rod about 10 mm diameter, the end slot being made by a hacksaw. Start the 'rip' with pliers (Fig. 13.7). Pick up the tag in the slot at the end of the rod and twist it, at the same time taking it round the cable; break into the ring and finish as with the nipper method (Fig. 13.9).

For light duty cables up to 4L1.5 in size the *Joistripper* tool is very efficient, it is quick and easy to use, and will take off more sheath than any other tool of its type, and is available from the manufacturers of MI cables, and their suppliers (Fig. 13.6).

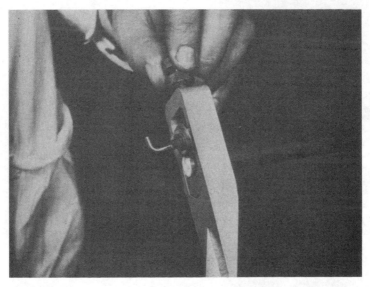

Fig. 13.10   *Large rotary stripping tool (BICC Pyrotenax)*

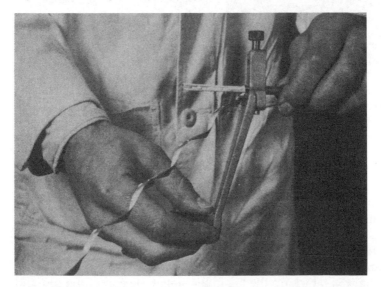

Fig. 13.11   *Small rotary BICC tool in use; sheath on open spiral (BICC Pyrotenax)*

For others cables, large or small rotary strippers can be used, these are also obtainable from the cable manufacturers (Figs. 13.10 and 13.11).

Sealing cable ends   The standard screw-on seal consists of a brass pot that is anchored to the cable sheath by means of a self-tapping thread. The pot is then filled with a sealing compound and the mouth of the pot closed by crimping home a stub cap or disc/sleeve assembly. The components necessary are determined by the conductor temperature likely to be encountered. They are as follows:

−80 to 105°C	Grey sealing compound, stub cap with PVC stub sleeving or fabric disc with headed PVC sleeving.
105 to 150°C	Grey sealing compound, stub cap with silicone elastomer glass sleeving or plastic disc with headed PTFE sleeving.
150 to 250°C	Glazing flux, ceramic cap with PTFE sleeving.
−20 to 100°C	Two-part epoxy putty, polypropylene disc, PTFE headed sleeving. (For use where explosive gases are likely to be present.)

Having ringed and stripped the sheath, slip the gland parts, if any, on to the cable.

To complete the screw-on seal, see that the conductors are clean and dry, engage the sealing pot square and fingertight on the sheath end, then tighten the pot with pliers or grips until the end of the cable sheath is level with the shoulder at the base of the pot. In general the cable should not project into the pot but a 1 or 2 mm projection is required for certain 250°C and increased safety seals. Alternatively the pot wrench can be used in conjunction with the gland body (Fig. 13.12).

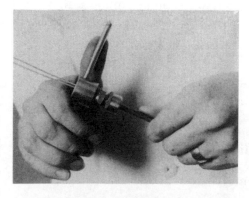

Fig. 13.12   *Sealing the cable end. A quick and accurate method of fitting the pot is by the use of a pot wrench (BICC Pyrotenax)*

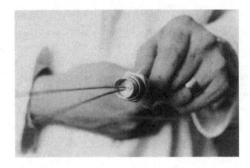

Fig. 13.13   *Examination of the pot interior prior to filling with compound (BICC Ltd)*

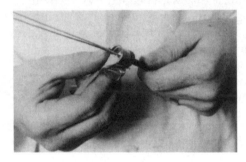

Fig. 13.14   *Overfilling the pot with compound. Use the plastic wrapping to prevent fingers coming into contact with the compound so as to ensure cleanliness of the seal (BICC Ltd)*

If the pot is difficult to screw on, moisten the sheath with an oil damped rag. To avoid slackness do not reverse the action. Examine the inside of the pot for cleanliness and metallic hairs, using a torch if the light is poor (Fig. 13.13). Test the pot for fit inside the gland. Set the conductors to register with the holes in the cap. Slip the cap and sleeving into position to test for fit, then withdraw slightly. Press compound into the pot until it is packed tight. The entry of the compound is effected by feeding in from one side of the pot only to prevent trapping air. To ensure internal cleanliness of the seal, use the plastic wrapping to prevent fingers from coming into contact with the compound (Fig. 13.14).

Next slide the stub cap over the conductors and press into the recess in the pot (Fig. 13.15). Finally, the pot must be crimped using a crimping tool (Fig. 13.16) and the termination completed by sliding insulated sleeves of the required length onto the conductors.

New types of seal are becoming available with the sealing compound supplied as an integral part of the seal (Fig. 13.5). These seals are easier to fit and information on their use may be obtained from cable manufacturers.

Information on through joints is given in Chapter 7.

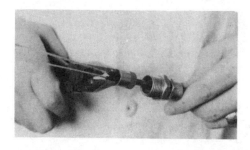

Fig. 13.15 *Using a pair of pliers to press the stub cap into the pot recess (BICC Ltd)*

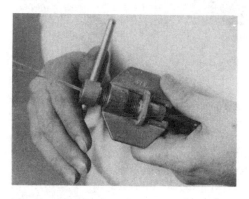

Fig. 13.16 *Securing the stub cap in position using a crimping tool which makes three indent crimps (BICC Pyrotenax)*

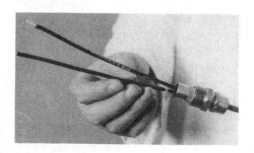

Fig. 13.17 *The stub sleeving is slid into position (BICC Pyrotenax)*

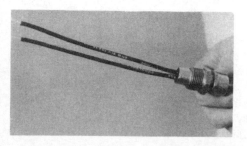

Fig. 13.18 *The completed termination (BICC Pyrotenax)*

Fig. 13.19   *Mineral insulated cables terminated in a fire alarm control panel enclosure. The gland nuts can be seen together with the sleeved conductors which will be trimmed to length when the fire alarm equipment is installed (British Telecom)*

Current ratings of cables

Owing to the heat-resisting properties of MI cables and to the fact that the magnesia insulation is a good conductor of heat, the current ratings of these cables are higher than those of VRI, PVC or even PI cables.

Multicore cables are not made larger than 25 mm^2, and therefore when heavier currents need to be carried it is necessary to use two or more single-core cables which are made in sizes up to 240 mm^2. Where single-core cables are run together their disposition should be arranged as shown in IEE Table 4A1. The current-carrying capacity of large single-core cables depends considerably upon their disposition.

IEE Tables 4J1A to 4J2B are for copper conductor MI cables. When these cables are run under conditions where they are not exposed to touch, they are rated to run at a comparatively high temperature and the current rating is considerably more than cables which are exposed to touch, or are covered with a PVC sheathing. For example a 150 mm^2 single-core cable is rated to carry 349 A if exposed to touch, but if not exposed to touch the same cable is rated to carry 485 A.

When an installation is designed to carry these higher currents, due regard must be paid to voltage drop, and also to the fact that the high temperature which is permissible in these cables might be transmitted to switchgear, and which might be affected by the conducted heat from the cable.

Some practical hints

These cables are supplied in coils, and every effort should be made to ensure that the coils retain their circular shape. They are frequently thrown off the delivery lorry and the impact flattens and hardens them. Before despatch the manufacturers anneal the cables so they are in a pliable state, but during transit and subsequent handling manipulation in excess of the manufacturers' recommendations will work harden the cable and could cause sheath fracture.

To measure the cable it should not be run out and recoiled as this tends to harden the cable. The best way is to measure the mean diameter of the coil and multiply by 3.14 which will give the approximate length of each turn in the coil.

Kinks or bends in the cable can best be removed by the use of a cable straightener. This is a device with pressure rollers that can be run backwards and forwards over the cable until the kinks are smoothed out.

The magnesium oxide insulation used in the cable has an affinity for moisture. There is, therefore, a need for temporary sealing during storage.

After sealing, an insulation test between conductors and to earth should be carried out, and this test should be repeated not less than 24 hours later. The second reading should have risen, and be at least 100 MΩ with a 500 V insulation tester.

As the conductors cannot be identified during the manufacturing process it is necessary to identify them after making off the seals. This can be done by fitting coloured sleeves or numbered markers onto the core. Correct identification can be checked by the use of a continuity tester.

Inductive loads

Switching of inductive loads can cause high voltage surges on 230 V and 400 V circuits, and these surges could cause damage to MI cables. Protection from these surges can be achieved by the use of inexpensive surge suppressors. The manufacturers of MI cables will be pleased to give advice on this matter.

14 Luminaires, switches, socket outlets and accessories

The final stage of electrical installation work is the fixing of accessories, such as ceiling roses, holders, switches, socket outlets and luminaires. This work requires experience and a thorough knowledge of the regulations which are applicable, because danger from shock frequently results from the use of incorrect accessories or due to accessories being wrongly connected.

IEE Regulation 132 lays down the requirements for the selection of equipment.

Ceiling roses

Ceiling roses may be of the 2-plate pattern and must also have an earth terminal (Fig. 12.2). The 3-plate type is used to enable the feed to be looped at the ceiling rose rather than to use an extra cable which would be needed to loop it at the switch.

For PVC-sheathed wiring it is possible to eliminate the need for joint boxes if 3-plate ceiling roses are employed (see Chapter 12). No ceiling rose may be used on a circuit having a voltage normally exceeding 250 V. Not more than two flexible cords may be connected to any one ceiling rose unless the latter is specially designed for multiple pendants.

Special 3- and 4-pin fittings rated at 2 or 6 A may be obtained and these can be installed where lighting fittings need to be removed or rearranged. The ability to remove lighting easily can assist in carrying out maintenance (Fig. 14.1). Although the fitting is a socket outlet, it cannot be used for supplying hand held equipment and under IEE Regulation 413-02-09 is exempt from the requirement of the 0.4 second disconnection time.

For the conduit system of wiring it is usual to fit ceiling roses which screw directly on to a standard conduit box, the box being fitted with an earth terminal.

Fig. 14.1   *A three-pin connector rated at 2 A designed to enable lighting to be easily removed and refitted (Ashley and Rock Ltd)*

Luminaires

Every luminaire or group of luminaires must be controlled by a switch or a socket outlet and plug, placed in a readily accessible position.

In damp situations, every luminaire shall be of the waterproof type, and in situations where there is likely to be flammable or explosive dust, vapour, or gas, the luminaires must be of the flameproof type.

Flexible cords

Flexible cords, if not properly installed and maintained, can become a common cause of fire and shock.

They must not be used for fixed wiring. Flexible cords must not be fixed where exposed to dampness or immediately below water pipes. They should be open to view throughout their entire length, except where passing through a ceiling when they must be protected with a properly bushed non-flammable tube. Flexible cords must never be fixed by means of insulated staples.

Where flexible cords support luminaires the maximum weight which may be supported is as follows:

$0.5\,mm^2$ $2\,kg$
$0.75\,mm^2$ $3\,kg$
$1.0\,mm^2$ $5\,kg$

In kitchens and sculleries, and in rooms with a fixed bath, flexible cords shall be of the PVC sheathed or an equally waterproof type.

When three-core flexible cords are used for fixed or portable fittings that have to be earthed, the colour of the cores shall be *brown* (connected to phase side), *blue* (connected to neutral or return), and *green/yellow* (connected to earth).

When four-core flexible cords are used for 3-phase appliances, the colours of the cores shall be *brown, black* and *grey* for the phases, *blue* for neutral, with *green/yellow* for the protective conductor.

Connections between flexible cords and cables shall be effected with an insulated connector, and this connector must be enclosed in a box or in part of a luminaire.

If an extension of a flexible cord is made with a flexible cord connector consisting of pins and sockets, the *sockets* must be fed from the supply, so that the exposed pins are not alive when disconnected from the sockets.

Where the temperature of the luminaire is likely to exceed 60°C, special heat-resisting flexible cords should be used for all tungsten luminaires, including pendants and enclosed type luminaires, the flexible cord should be insulated with butyl rubber or silicone rubber. Ordinary PVC insulated cords are not likely to stand up to the heat given off by tungsten lamps. Flexible cords feeding electric heaters must also have heatproof insulation such as butyl or silicone rubber.

Where extra high temperatures are likely to be encountered it is advisable to consult a cable manufacturer before deciding on the type of flexible cord to use.

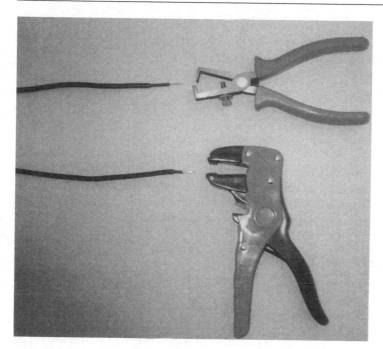

Fig. 14.2 *The best way of stripping insulation from small cables or flexible cords is to use a stripping tool. A number of types are available and two of them are illustrated here. The type at the top has a screw adjustment and is preferred by some electricians. Others, such as the example shown below, grip the insulation and automatically set the cutters to the correct gap to strip the insulation*

Flexible cords used in workshops and other places subjected to risk of mechanical damage shall be PVC sheathed or armoured.

All flexible cords used for portable appliances shall be of the sheathed circular type and, therefore, twisted cords must not be used for portable handlamps, floor and table lamps, etc.

The definition of a 'flexible cord' is 'A flexible cable in which the cross sectional area of each conductor does not exceed 4 mm²'. Larger flexible conductors are known as 'flexible cables'.

All flexible cords should be frequently inspected, especially at the point where they enter lampholders and other accessories, and renewed if found to be unsatisfactory.

Lampholders

Insulated lampholders should be used wherever possible. Lampholders fitted with switches must be controlled by a fixed switch or socket outlet in the same room.

The outer screwed contact of Edison screw-type lampholders must always be connected to the neutral of the supply. Small Edison screw lampholders must have a protective device not exceeding 6 A, but the larger sizes may have a protective device not exceeding 16 A.

The small BC lampholder must have a protective device not exceeding 6 A, and for the larger BC lampholders the protective device must not exceed 16 A (IEE Table 55B).

No lampholder may be used on circuits exceeding 250 V (IEE Regulation 553-03-02), and all metal lampholders must have an earth terminal. In bathrooms, and other positions where there are stone floors or exposed extraneous conductive parts, lampholders should be fitted with insulated skirts to prevent inadvertent contact with live pins when a lamp is being removed or replaced.

Socket outlets and plugs

The 13 A socket outlet with fused plug made to BS 1362 and BS 1363 is in general use for domestic and office premises. The 13 A socket outlet is also extensively used in industrial premises. Socket outlets to BS 196 are also used for circuits not exceeding 250 V, and are made in ratings of 5 A, 15 A and 30 A.

Other industrial type socket outlets are covered by BS EN 60309, and these include single-phase and three-phase with ratings up to 125 A.

Details of the ratings and circuiting of these various types of socket outlets are given in Chapter 5.

The Low Voltage Electrical Equipment (Safety) Regulations 1989 require equipment to be safe. This implies that any part intended to be electrified is not to be capable of being touched with a finger, and this includes a child's finger. Thus the live pins of plugs should be partly shrouded so that when the plug is in the process of being inserted even the smallest finger cannot make contact with live metal.

BS 1363, clause 4-2-2 requires that socket outlets shall be provided by a screen which automatically covers the live contacts when the plug is withdrawn.

When installing socket outlets the cables must be connected to the correct terminals, which are:

brown wire (phase or outer conductor) to terminal marked L
blue wire (neutral or middle conductor) to terminal marked N
yellow/green earth wire to terminal marked E.

Flexible cords connected to plugs shall be brown (phase) blue (neutral) and yellow/green (earth).

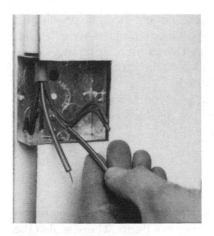

Fig. 14.3 *Wiring 13A socket outlet on ring circuit (1). Draw at least 120–150 mm of each cable into the box, and then strip back the outer sheath of each cable to within some 20 mm of the knock-out of cable entry. The ends of the conductors should be bared and the tails shaped to conform to the terminal positions in the socket outlet. Suitable green/yellow earth sleeves should be cut and slid onto the protective conductor*

Fig. 14.4 *Wiring 13A socket outlet (2). Insert conductors fully into the appropriate terminal and tighten the terminal screw (socket outlets with tunnel type terminals should be used for preference as these terminals enable maximum and uniform pressure to be applied on up to two main circuit cables and one spur cable)*

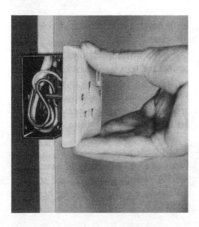

Fig. 14.5 *Wiring 13A socket outlet (3). After all terminal connections have been made the slack cables should be carefully disposed to avoid cramping. Finally the socket outlet may be pushed gently into the box and secured by the two fixing screws. (To show how neatly the cables lie, even within a shallow box, our illustration is of a box with side cut away) (Crabtree Electrical Industries Ltd)*

If wrong connections are made to socket outlets it may be possible for a person to receive a shock from an appliance when it is switched off.

Socket outlet adaptors which enable two or more appliances to be connected to a single socket should contain fuses to prevent the socket outlet from becoming overloaded.

Socket outlets installed in old people's homes and in domestic premises likely to be occupied by old or disabled people, should be installed at not less than 1 m from floor level.

Switches

There are various types of switches available, the most common being the 6 A switch which is used to control lights. There is also the 15 A switch for circuits carrying heavier currents.

For a.c. circuits the micro-gap switch is being extensively used; it is much smaller than the older type and more satisfactory for breaking inductive loads.

Quick-make and slow-break switches are recommended for a.c.

A quick-break switch connected to an a.c. supply and loaded near to its capacity will tend to break down to earth when used to switch off an inductive load (such as fluorescent lamps).

In a room containing a fixed bath, switches are only permitted outside specified zones, ensuring they are out of reach of a person in the bath. Insulated cord operated ceiling switches may be fitted in zones 1 and 2 (Fig. 14.6).

The correct method of mounting switches for the various wiring systems is dealt with in the sections which cover these systems.

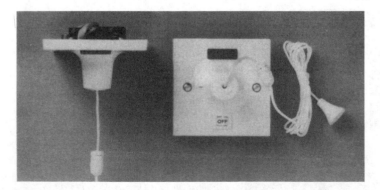

Fig. 14.6   *30 A double pole ceiling switch (MK Ltd)*

Fig. 14.7   *Polycarbonate splashproof sockets or switches are available for use in damp or dusty situations (MK Ltd)*

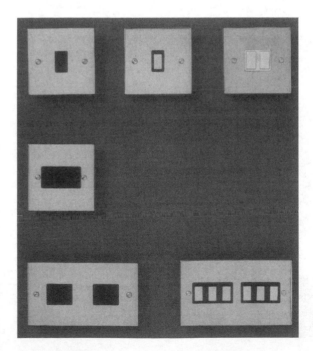

Fig. 14.8   *A range of different types of flush-mounted switches (Contactum Ltd)*

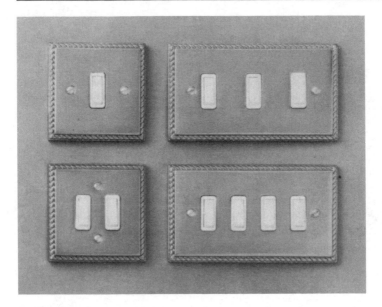

Fig. 14.9   A range of decorative flush-mounted switches (MK Ltd)

Fig. 14.10   A range of Metalclad switches and 13 A socket outlets (Temby Industries Ltd)

All single-pole switches shall be fitted in the same conductor throughout the installation, which shall be the phase conductor of the supply.

In damp situations, every switch shall be of the waterproof type with suitable screwed entries or glands to prevent moisture entering the switch or socket (Fig. 14.7).

To prevent condensed moisture from collecting inside a watertight switchbox, a very small hole should be drilled in the lowest part of the box to enable the moisture to drain away.

Flameproof switches must be fitted in all positions exposed to flammable or explosive dust, vapour or gas.

15 *Earthing*

The object of earthing a consumer's installation is to ensure that all exposed conductive parts and extraneous conductive parts associated with electrical installations are at, or near, earth potential.

Earthing conductors and protective conductors need to satisfy two main requirements, namely to be strong enough to withstand any mechanical damage which is likely to occur, and also to be of sufficiently low impedance to meet the need to carry any earth fault currents without danger.

The supply authority connect the neutral point of their transformer to earth, so as to limit the value of the phase voltage to earth. The consumer's earthing system must be so arranged to ensure that in the event of an earth fault of negligible impedance, the fault current shall not be sustained so as to cause danger. The protective devices in the circuit (e.g. fuses or circuit breakers) must operate so as to disconnect the fault within the maximum times specified in the regulations. The protective conductors and earthing system must be arranged so as to ensure this happens.

IEE Regulations Section 413 and Chapter 54 deal with the design aspects for earthing and the provision of protective conductors, and a number of points require consideration when dealing with this part of the installation design.

The terms used and types of protective conductor are illustrated in Fig. 15.1. The types of protective conductor shown are as follows:

(1) **Earthing conductor** This connects the main earthing terminal with the means of earthing which may be an earth electrode buried in the ground for a TT system, or where a TN system is in use, another means of earthing such as the supply authority terminal.

(2) **Circuit protective conductor** These are run for each circuit and may comprise a separate conductor, be incorporated in the cable for the circuit concerned or be the metal conduit or

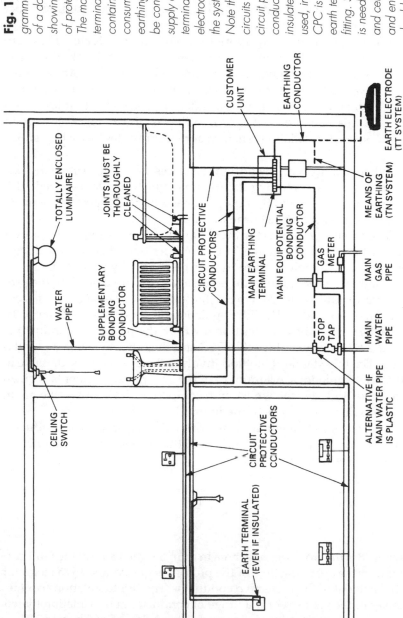

Fig. 15.1 A diagrammatic representation of a domestic installation showing the main types of protective conductor. The main earthing terminal is normally contained in the consumer unit, and the earthing conductor will be connected to the supply authority's earthing terminal or an earth electrode depending on the system of supply. Note that all the lighting circuits must have a circuit protective conductor even if insulated fittings are used, in which case the CPC is terminated in an earth terminal in the fitting. Special bonding is needed in bathrooms, and ceiling light switches and enclosed luminaries should be used

Labels in figure:

TOTALLY ENCLOSED LUMINAIRE

JOINTS MUST BE THOROUGHLY CLEANED

CUSTOMER UNIT

EARTHING CONDUCTOR

WATER PIPE

SUPPLEMENTARY BONDING CONDUCTOR

CIRCUIT PROTECTIVE CONDUCTORS

MAIN EARTHING TERMINAL

MAIN EQUIPOTENTIAL BONDING CONDUCTOR

GAS METER

MAIN GAS PIPE

MEANS OF EARTHING (TN SYSTEM)

EARTH ELECTRODE (TT SYSTEM)

STOP TAP

MAIN WATER PIPE

CEILING SWITCH

ALTERNATIVE IF MAIN WATER PIPE IS PLASTIC

EARTH TERMINAL (EVEN IF INSULATED)

CIRCUIT PROTECTIVE CONDUCTORS

cable sheath in, for example, the case of mineral insulated cables.

(3) **The main equipotential bonding conductor** These connect the main earthing terminal with the main service metal pipes such as water and gas, and with any exposed building structural steelwork, ventilation ducting, etc.

(4) **Supplementary equipotential bonding conductor** These are needed in locations where there is increased risk of electric shock (IEE Regulations part 6 and 471-08-01). An example would be a bathroom where bonding is required to connect metal parts such as pipes, radiators, accessible building parts baths and shower trays. Bonding is also required in certain special installations such as agricultural sites. See Chapter 6.

When dealing with the design it is necessary to determine the size of the various protective conductors which are to be used. IEE Regulation 543-01 and Section 547 deal with this aspect and a number of points need to be borne in mind.

The protective conductor must have sufficient strength to protect against mechanical damage. The minimum size where the conductor is not part of a cable is $4\,\text{mm}^2$ unless mechanically protected.

Thermal considerations are necessary to ensure that when the protective conductor is carrying a fault current, damage to adjacent insulation is avoided. The IEE Regulations give two 'standard methods' of determining the cross sectional size of the protective conductors. One is by the use of IEE Table 54G and the other by using a calculation. The first standard method involves the use of a look-up table in the IEE Regulations. A summary of the general requirements is given in Table 15.1 which, whilst not comprehensive, shows the main points which need to be borne in mind.

The calculation method of determination of protective conductor size is with the use of a formula, and this is given in IEE Regulation 543-01-03 as follows:

$$S = \frac{\sqrt{I^2 t}}{k}$$

where S is the cross sectional area in mm^2, I is the value of the fault current, t is the operating time of the protective device, and k is a factor for the material of the protective conductor. The IEE Regulations contain tables of the value of k for a range of conductor and insulation materials. The use of the formula is more fully explained in Chapter 2.

Table 15.1 *General requirements for determination of the size of the protective conductors*

Type of protective conductor	Minimum size	Method of calculation
Earthing conductor	$16\,mm^2$ or $25\,mm^2$ unless protected (see IEE Table 54A)	'Standard method' (1)
Main equipotential bonding conductor	Min $6\,mm^2$ (max $25\,mm^2$) (2)	Not less than half of the earthing conductor's cross-sectional area
Circuit protective conductor	$1.0\,mm^2$ if forming part of a cable	'Standard method' (1)
Supplementary bonding conductor	$4\,mm^2$ unless mechanically protected. Otherwise $2.5\,mm^2$ or dependent on size of CPC.	Calculation size specified in IEE Regulation 413-02-28 with minimum sizes in 547-03

Notes:
(1) The 'standard method' is given in IEE Regulation 543-01 and comprises either use of a calculation using a formula, or reference to a look-up table (IEE Table 54G).
(2) For PME systems, Min $10\,mm^2$, Max $50\,mm^2$ depending upon the size of the supply neutral. (Use IEE Table 54H.)

Table 15.2 *Minimum cross-sectional area of protective conductors in relation to the area of associated phase conductors (Extract from IEE Table 54G)*

Cross-sectional area of phase conductor (mm^2)	Minimum cross-sectional area of the corresponding protective conductor (mm^2) of same material as phase conductor
not exceeding 16	same size as phase conductor
exceeding 16 but not exceeding 35	16
exceeding 35	half the size of the phase conductor

A second method of determining the protective conductor size is to use IEE Table 54G. Part of the IEE Table is reproduced in Table 15.2. Although this method is the simpler to apply, it can result in the provision of a protective conductor larger than is strictly necessary, and this will, of course, increase the cost of the installation. Where it is intended to use 'two core and CPC' cables, the protective conductor incorporated within them is generally smaller than the live conductors, and as a result IEE Table 54G cannot be applied.

IEE Table 54G gives additional useful information relevant where the protective conductor is of a different material to the phase conductor, such as could be encountered with steel wire armoured cable.

Note that the first (calculation) method has to be used when the size of the phase conductors has been determined by the expected short circuit current in the circuit and the earth fault current is expected to be less than the short circuit current. The total earth fault loop impedance needs to be checked to ensure that shock prevention measures for indirect contact are achieved. Information on this is given in Chapter 2 and it is necessary to ensure that the disconnection times and earth fault loop impedances using IEE Tables 41B1, 41B2, 41C or 41D are met.

Gas and water pipes and other extraneous conductive parts as mentioned above must not be used as an earth electrode of any installation. The consumer's earth terminal may be a connection to the supply undertaking's earth point (if provided by them) otherwise an independent earth electrode must be provided, consisting of buried copper rods, tapes, pipes, or plates, etc., as detailed in IEE Regulation 542-02-01.

Most types of armoured multicore cables rely upon their metal sheathing or armouring to serve as a protective conductor, but it must not be assumed that all multicore armoured cables have armouring of sufficiently low impedance (especially in long runs of cable) to permit sufficient fault current to flow to operate the protective device. In some cases it may be necessary to provide an additional protective conductor in the form of an earth tape in parallel with the cable or an additional core in the cable.

Metal conduit and trunking are suitable to serve as protective conductors, providing that all joints are properly made, and their conductance is at least equal to the values required in IEE Regulation 543-01.

Flexible or pliable conduit shall not be used as a protective conductor (IEE Regulation 543-02-01), and where final connections are made to motors by means of flexible conduit, a separate circuit protective conductor should be installed within the flexible conduit to bond the motor frame to an earthing terminal on the rigid conduit or starter.

Steel catenary wires may be used as protective conductors providing their conductance is satisfactory. It should be remembered that the resistivity of steel is approximately eight times that of copper.

Additional information may be obtained by reference to British Standard Code of Practice BS 7430 which gives a lot of detailed information on earthing requirements and methods.

Fig. 15.2 *The main earthing terminal in an industrial installation. The earthing conductor is at the bottom of the picture, and flat copper strip is used for some of the bonding. Circuit protective conductors can also be seen (British Telecom)*

Fig. 15.3 *Another example of an industrial earthing location. The connections are located where inspection can readily be achieved. (W.T. Parker Ltd)*

In a room containing a fixed bath or shower, supplementary equipotential bonding shall be provided between simultaneously accessible exposed conductive parts of equipment, between exposed conductive parts and between extraneous conductive parts (IEE Regulation 601-04-01). This means that all exposed metalwork, such as pipes, must be bonded together (Fig. 15.1).

Fig. 15.4   *A commercial location where a cable riser is constructed from cable basket and carries cables for different types of circuit. Note the equipotential bonding conductors. (W.T. Parker Ltd)*

Consideration should be given to bonding in vulnerable situations such as kitchens, laundries, milking parlours, laboratories, etc., where persons or animals may be exposed to exceptional risks of electric shock. In these situations, consideration should be given to using residual current circuit breakers.

IEE Regulation 601-04-01 requires that the earth arrangements in bathrooms shall be such that in the event of an earth fault disconnection occurs within 0.4 s for all types of equipment.

16 *Inspection and testing*

Inspection and testing of electrical installations are dealt with in Part 7 of the IEE Regulations.

Chapter 71, Regulation 711-01-01 states that 'Every installation shall, during erection and/or on completion and before being put into service, be inspected and tested to verify, as far as is reasonably practicable, that the requirements of the Regulations have been met'.

Inspection

IEE Section 712 lists the areas of the installation which must be inspected and this should be carried out before the testing work is started. The inspection can of course be undertaken during the erection of the equipment, and in some cases this is the only practical way of carrying out this work. The person carrying out the inspection and testing work will need to have an understanding of the design concept used in the installation, and will thus need to have available the data from the design phase relating to the assessment of general characteristics. It is also necessary to have available charts or diagrams showing the arrangements of each circuit, as listed in IEE Regulation 514-09-01.

IEE Regulation 712-01-03 lists items which should be covered by the inspection work, and these include:

electrical connections
identification of conductors
safe routing of cables
conductors are selected in accordance with the design
that single pole devices are connected in the phase conductor
correct connection of sockets, accessories and equipment
presence of fire barriers
appropriate insulation of conductors
presence of protective conductors   (*Continued overpage*)

appropriate isolators and switches
methods of protection against electric shock
prevention of mutual detrimental influence
undervoltage protection
danger notices and labelling of circuits, fuses, etc.
access to switchgear is adequate.

Special attention should be paid to flexible cords which are connected to portable appliances, especially where they enter plugs and appliances.

In old installations the insulation of cables where they enter distribution boards and switchfuses should be carefully examined, as very often the insulation becomes damaged and brittle due to heat.

All protective devices and fuselinks should be checked to ensure they are of the correct rating to protect the circuit cables which they control.

Testing

IEE Regulations 713-02 to 713-12 detail the standard methods of testing required. The tests should be as follows, and should be carried out in the sequence indicated:

continuity of protective conductors
continuity of final circuit ring conductors
insulation resistance
insulation of site-built assemblies
protection by separation of circuits
protection by barriers or enclosures
insulation of non-conducting floors and walls
polarity
earth electrode resistance
earth fault loop impedance
prospective fault current
functional tests including the operation of residual current devices.

The methods of making these tests are detailed in IEE Chapter 71.

The Health and Safety Executive has issued a guide on Electrical Testing (HS(G) 13 HMSO) which gives advice on precautions which should be taken when testing live installations.

The guide mentions that many accidents occur when making these tests. It recommends that bare ends of test probes should not exceed 2 to 3 mm of bare metal, and that metal lampholders should never be used for test lamps.

Some hints, based upon practical experience, are given here to supplement the advice contained in the Regulations.

Continuity tests

To test the continuity of ring final circuit conductors, a digital ohm-meter or multimeter set to 'ohms' range should be used. The ends of the ring circuit conductors are separated and the resistance values noted for each of the live conductors and for the protective conductor. The ring circuit is then reconnected and a further resistance measurement taken for each conductor between the distribution board and the appropriate pin of the outlet nearest to the mid-point of the ring. The value obtained should be approximately one quarter of the value of the first reading for each conductor. The test lead needed to carry out the second part of this test will be quite long, and it will be necessary to determine its resistance and deduct the figure from the readings obtained to obtain a valid result.

An alternative method of testing a ring circuit avoids the use of a long test lead. It is initially necessary to determine which ends are which for the installed ring circuit. This is done by shorting across the phase and neutral conductors of the first or last socket outlet on the ring, and applying an ohmmeter to the cable ends at the distribution board (see Fig. 16.1a). If the readings of the test meter are different in position A than in position B the pairs are matched correctly and the test may be continued. If the readings are the same in position A and in position B, the short and long sides of the ring are linked, and the wrong pairs have been selected. The test is unacceptable.

The next step is to remove the short circuit from the first or last socket outlet on the ring. Then short together the live conductor of one of the pairs of cables and the neutral conductor of the other. Also short together the remaining pair of cables (see Fig. 16.1b).

Next the test instrument is connected to each socket outlet on the ring in turn. The resistance reading in each position should be identical, and if it is, the continuity is proved. If one of the readings is different, the socket outlet either is connected as a spur to the ring circuit or is a socket outlet on a different ring.

Special instruments are now available for checking the resistance of the CPC or of metal conduit and trunking where it is used as part of the protective conductor. The instrument operates by applying a current at extra low voltage to the section of conduit or trunking

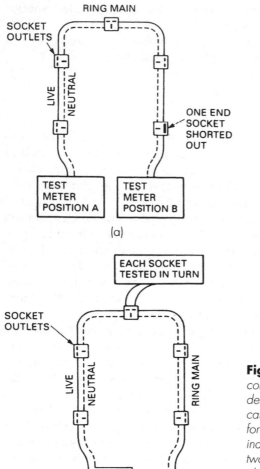

Fig. 16.1 *Testing the continuity of a ring circuit as described in the text: (a) indicates the end socket shorted out for the initial test to identify the individual cables; (b) shows part two of the test, applying the ohmmeter to each socket in turn and comparing the resistance readings. All the resistances must be identical to show that continuity is proved*

connected, and gives a reading of the continuity in ohms. The characteristics of such an instrument are included in IEE Regulation 713-02-01.

Insulation resistance

Insulation tests should be made with a portable insulation tester, comprising a hand-operated generator or a battery-operated

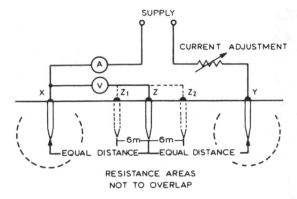

Fig. 16.2 *Measurement of earth electrode resistance. X – earth electrode under test, disconnected from all other sources of supply; Y – auxiliary earth electrode: Z – second auxiliary earth electrode; Z_1 – alternative position of Z for check measurement; Z_2 – further alternative position of Z for check measurement. If the tests are made at power frequency the source of the current used for the test shall be isolated from the mains supply (e.g. by a double-wound transformer), and in any event the earth electrode X under test shall be disconnected from all sources of supply other than that used for testing*

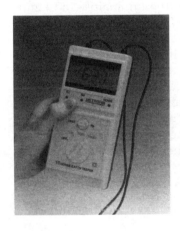

Fig. 16.3 *A hand-held earth/ground tester. It will measure resistance from 0.01 to 2000 Ω (Edgcumbe Instruments Ltd)*

instrument, with a scale reading in ohms. The voltage of the instrument should be

250 V d.c. for extra low voltage circuits
500 V d.c. for low voltage circuits up to 500 V, and
1000 V d.c. for low voltage circuits between 500 V and 1000 V.

Suitable instruments for making these tests are shown in Figs. 16.4 and 16.5.

Fig. 16.4 *A 250/500/1000V insulation and continuity tester, with digital display (AVO International)*

The main test should be made before the luminaires and lamps are installed, but with all fuses inserted, all switches on, and the conductors of both poles connected together, and with the supply switched off. This test will be between all conductors bunched, and earth. The result of the test should be not less than $0.5\,M\,\Omega$. Particular attention should be given to the presence of electronic devices connected to the installation, and such devices should be isolated so that they are not damaged by the test voltage.

Another test is between phase and neutral conductors, after all lamps have been removed, and all switches in the 'on' position. This test shall produce a reading of not less than $0.5\,M\,\Omega$. If a reading lower than $0.5\,M\,\Omega$ is obtained then steps must be taken to trace and rectify the fault.

Polarity

Verification of polarity is to ensure that all single pole switches, including switches on socket outlets, are connected to the live, or outer conductor, of the supply. For lighting circuits all lamps should be removed, and, with the current switched on, a 15W test lamp should be connected between the live terminal of the switch to earth. If the polarity is correct the lamp will fully light, if there is no light the polarity is incorrect.

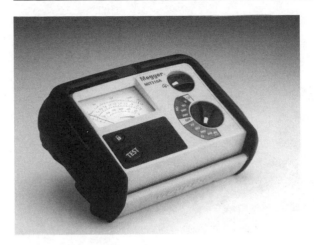

Fig. 16.5 *A battery-operated insulation resistance test instrument operates at 250, 500 or 1000 V (Megger Ltd)*

For socket outlets a test should be made with a 15 W test lamp connected to the L and E terminals of a plug. When the plug is inserted into a socket outlet, if the socket outlet is correctly connected, the test lamp will fully light; in a switched socket outlet it should light only when the switch on the socket outlet is in the *on* position. This test will also indicate if the earth connection is properly made.

Another polarity test is to ensure that all Edison-screw lampholders are correctly connected. The outer ring of the lampholder must be connected to the neutral or earthed conductor. This test can be made by connecting a 15 W test lamp between the outer ring of the lampholder and earth. There should be no light when so connected, but the lamp should light when connected between the inner contact and earth.

When making this test care must be taken to avoid accidentally shortcircuiting outer and inner lampholder contacts.

Earth electrode resistance

The test should be carried out with an earth tester similar to that shown in Fig. 16.3, or a more modern digital equivalent. An alternating current is passed between points X and Y and an additional earth spike Z is placed successively at points Z_1, Z_2, etc. Voltage drops between X and Z, and Z and Y are obtained for successive positions of Z and the earth electrode resistance is calculated and checked from the voltage drop and current flowing.

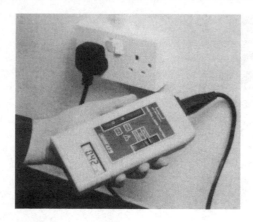

Fig. 16.6 *A digital loop
tester in use measuring the earth
fault loop impedance at a
socket outlet (AVO International)*

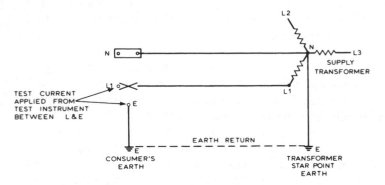

Fig. 16.7 *Earth fault loop impedance test measures the impedance in the
line-earth loop which comprises the following parts: the circuit protective conductor; the consumer's earthing terminal and earthing conductor; the earth return
path through the general mass of earth; the supply transformer earth; the neutral
point of the supply transformer and winding; the phase conductor*

Earth fault loop impedance

Tests for earth fault loop impedance should be made with an instrument such as that shown in Fig. 16.6.

The object of this test (Fig. 16.7) is to ensure that the phase earth
loop impedance of the circuit is appropriate to the rating and type of
protective device as specified by the IEE Wiring Regulations and
thus ensure the circuit will disconnect within the correct time. If a
fault did not result in the fuse or circuit breaker disconnecting in the
correct time, a very dangerous state of affairs could exist, and it is
important that this test be made and acted upon.

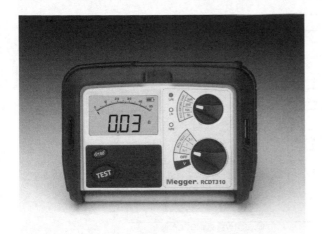

Fig. 16.8   *A residual current circuit breaker test set (Megger Ltd)*

Testing residual current circuit breakers

The operation and use of residual current circuit breakers were described in Chapter 2. Test instruments can be obtained which are designed to carry out tests of residual current devices and the instrument is connected to the load side of the device, the loads themselves being disconnected. The test instrument simulates a fault so that a residual current flows, and then measures the response time of the residual current device, generally displaying the result in milliseconds. Instruments for this purpose are illustrated in Figs. 16.8 and 16.10. Residual current devices incorporate an integral test button and the effectiveness of this should also be tested.

The installation of voltage operated earth leakage circuit breakers is not now permitted by IEE Regulations. However, their use may be encountered in existing installations, and details of a test method suitable for them is given in Fig. 16.9. Voltage operated devices have a number of disadvantages and if any doubt exists as to their performance, they should be replaced by residual current circuit breakers.

Completion Certificates and Inspection Report Forms

A Completion Certificate and an Inspection Report Form must be given by the person responsible for the construction of the installation, or alteration thereto, or by an authorised person acting for them.

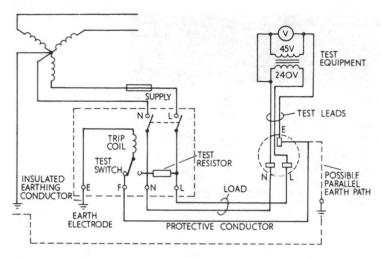

Fig. 16.9   *Voltage operated earth leakage circuit breakers are not now permitted by IEE Regulations. However their use may be encountered in existing installations, and testing of them may be carried out using this circuit. A test voltage not exceeding 50 V a.c. obtained from a double wound transformer of at least 750 VA is connected as shown. If satisfactory the circuit breaker will trip instantaneously*

Details of these certificates are given in IEE Appendix 6. The person who carries out any installation work assumes a very great responsibility in ensuring that the certificates are completed and that their terms are complied with in every respect.

Any loss or damage incurred due to any neglect on the part of the person responsible for the installation might well involve claims for heavy damages.

Notice of re-inspection and testing

IEE Regulation 514-12 states that a notice, of such durable material as to be likely to remain easily legible throughout the life of the installation, shall be fixed in a prominent position at or near the main distribution board upon completion of the work. It shall be inscribed as follows, in characters not smaller than those illustrated in Fig. 16.11. The determination of the frequency of periodic inspection is covered by IEE Regulation 732-01-01. No specific period is laid down, and an assessment needs to be made as to the use of the installation, the likely frequency of maintenance, and the possible external influences likely to be encountered. The person carrying out the inspection and

Fig. 16.10 *A digital residual current device tester. Devices with ratings from 6 to 500 mA may be tested (Edgcumbe Instruments Ltd)*

IMPORTANT

This installation should be periodically inspected and tested, and a report on its condition obtained, as prescribed in BS 7671 Requirements for Electrical Installations published by the Institution of Electrical Engineers.

Date of last inspection ....

Recommended date of next inspection ....

Fig. 16.11 *Wording specified by the IEE Regulations for the periodic inspection notice*

testing, and completing the inspection certificate needs to take account of these issues. In the absence of other local or national regulations, a maximum period of five years would be applied, with shorter periods where appropriate.

Index

trunking systems, 261–6
 plastic, 264–6
 steel, 261–3
Undervoltage, protection against,
 37–8
Use of buildings, 19

Vertical:
 rising mains, 269–70
 trunking, 251, 252
Voltage:
 drop, 56–8, 75–6, 96, 98
 exceeding 230V a.c., 127–8
 extra-low, 184, 197
 See also Low voltage

Wall outlets, conduit runs, 240–1
Warning, underfloor, 164–6
Warning label:
 supplies exceeding 230V a.c., 128
 wiring colours, 130
Water heating, off-peak, 163
Water pipes, 223–5, 320
Wattless current, 135–6

Wiring:
 colour identification, 129–33
 concealed, 288–92
 emergency supplies, 175–6, 251
 IEE Regulations BS 7671, 5, 6–12,
 16, 18
 tables, 93–102
 low voltage, 184
 sunk conduits, 242
 surface, 285–7
 switch wires, 130–2
 systems, 181–2
 all-insulated, 284–92
 choice of, 182–3
 See also Cable
Wooden floors, conduit, 237, 238
Worked example, 69–82
Working:
 dead, 15
 live, 15
Workmanship, 184

XLPE cables, 188, 273, 274